*About the Marine Sanctuaries Conservation Series*

*The National Oceanic and Atmospheric Administration's National Ocean Service (NOS) administers the National Marine Sanctuary Program (NMSP). Its mission is to identify, designate, protect and manage the ecological, recreational, research, educational, historical, and aesthetic resources and qualities of nationally significant coastal and marine areas. The existing marine sanctuaries differ widely in their natural and historical resources and include nearshore and open ocean areas ranging in size from less than one to over 5,000 square miles. Protected habitats include rocky coasts, kelp forests, coral reefs, sea grass beds, estuarine habitats, hard and soft bottom habitats, segments of whale migration routes, and shipwrecks.*

*Because of considerable differences in settings, resources, and threats, each marine sanctuary has a tailored management plan. Conservation, education, research, monitoring and enforcement programs vary accordingly. The integration of these programs is fundamental to marine protected area management. The Marine Sanctuaries Conservation Series reflects and supports this integration by providing a forum for publication and discussion of the complex issues currently facing the National Marine Sanctuary Program. Topics of published reports vary substantially and may include descriptions of educational programs, discussions on resource management issues, and results of scientific research and monitoring projects. The series facilitates integration of natural sciences, socioeconomic and cultural sciences, education, and policy development to accomplish the diverse needs of NOAA's resource protection mandate.*

# Olympic Coast National Marine Sanctuary Habitat Mapping: Survey report and classification of side scan sonar data from surveys HMPR-114-2004-02 and HMPR-116-2005-01

Steven S. Intelmann[1], and Guy R. Cochrane[2]

[1]Olympic Coast National Marine Sanctuary, NOAA
[2]Coastal and Marine Geology Program, USGS

U.S. Department of Commerce
Carlos M. Gutierrez, Secretary

National Oceanic and Atmospheric Administration
VADM Conrad C. Lautenbacher, Jr. (USN-ret.)
Under Secretary of Commerce for Oceans and Atmosphere

National Ocean Service
John H. Dunnigan, Assistant Administrator

National Marine Sanctuary Program
Daniel J. Basta, Director

Silver Spring, Maryland
September 2006

## DISCLAIMER

Report content does not necessarily reflect the views and policies of the National Marine Sanctuary Program or the National Oceanic and Atmospheric Administration, nor does the mention of trade names or commercial products constitute endorsement or recommendation for use.

## REPORT AVAILABILITY

Electronic copies of this report are available from the National Marine Sanctuary Program web site at *www.sanctuaries.nos.noaa.gov*. Hard copies are available from the following address:

National Oceanic and Atmospheric Administration
National Marine Sanctuary Program
SSMC4, N/ORM62
1305 East-West Highway
Silver Spring, MD 20910

## COVER

The cover features a sonar waterfall display showing an area where the towfish unfortunately slammed into a rock wall during the HMPR-114-2004-02 survey. The red line represents the towfish altitude up to the point of impact when sonar packets ceased to be recorded. This image represents an excellent example of what sonar operators hope never to see.

## SUGGESTED CITATION

Intelmann, S.S. and G.R. Cochrane. 2006. Olympic Coast National Marine Sanctuary Habitat Mapping: Survey report and classification of side scan sonar data from surveys HMPR-114-2004-02 and HMPR-116-2005-01. Marine Sanctuaries Conservation Series MSD-06-07. U.S. Department of Commerce, National Oceanic and Atmospheric Administration, National Marine Sanctuary Program, Silver Spring, MD. 35 pp.

## CONTACT

Steven S. Intelmann
Habitat Mapping Specialist
NOAA/National Marine Sanctuary Program
N/ORM 6X26
115 E. Railroad Avenue, Suite 301
Port Angeles, WA 98362
(360) 457-6622 X22
*steve.intelmann@noaa.gov*

# ABSTRACT

The Olympic Coast National Marine Sanctuary (OCNMS) continues to invest significant resources into seafloor mapping activities along Washington's outer coast (Intelmann and Cochrane 2006; Intelmann et al. 2006; Intelmann 2006). Results from these annual mapping efforts offer a snapshot of current ground conditions, help to guide research and management activities, and provide a baseline for assessing the impacts of various threats to important habitat. During the months of August 2004 and May and July 2005, we used side scan sonar to image several regions of the sea floor in the northern OCNMS, and the data were mosaicked at 1-meter pixel resolution. Video from a towed camera sled, bathymetry data, sedimentary samples and side scan sonar mapping were integrated to describe geological and biological aspects of habitat. Polygon features were created and attributed with a hierarchical deep-water marine benthic classification scheme (Greene et al. 1999). For three small areas that were mapped with both side scan sonar and multibeam echosounder, we made a comparison of output from the classified images indicating little difference in results between the two methods. With these considerations, backscatter derived from multibeam bathymetry is currently a cost-efficient and safe method for seabed imaging in the shallow (<30 meters) rocky waters of OCNMS. The image quality is sufficient for classification purposes, the associated depths provide further descriptive value and risks to gear are minimized. In shallow waters (<30 meters) which do not have a high incidence of dangerous rock pinnacles, a towed multi-beam side scan sonar could provide a better option for obtaining seafloor imagery due to the high rate of acquisition speed and high image quality, however the high probability of losing or damaging such a costly system when deployed as a towed configuration in the extremely rugose nearshore zones within OCNMS is a financially risky proposition. The development of newer technologies such as intereferometric multibeam systems and bathymetric side scan systems could also provide great potential for mapping these nearshore rocky areas as they allow for high speed data acquisition, produce precisely geo-referenced side scan imagery to bathymetry, and do not experience the angular depth dependency associated with multibeam echosounders allowing larger range scales to be used in shallower water. As such, further investigation of these systems is needed to assess their efficiency and utility in these environments compared to traditional side scan sonar and multibeam bathymetry.

# KEY WORDS

Benthic, habitat mapping, sediment classification, side scan sonar, multibeam echosounder, textural analysis, Olympic Coast National Marine Sanctuary, essential fish habitat, groundtruthing

i

# TABLE OF CONTENTS

## LIST OF FIGURES AND TABLES

# INTRODUCTION

The Olympic Coast National Marine Sanctuary (OCNMS) continues to invest significant resources into seafloor mapping activities along Washington's outer coast (Intelmann and Cochrane 2006; Intelmann et al. 2006; Intelmann 2006). Results from these annual mapping efforts offer a snapshot of current ground conditions, help to guide research and management activities, and provide a baseline for assessing the impacts of various threats to important habitat.

During the months of August 2004 and May/July 2005, we used side scan sonar to image several regions of the sea floor in the northern OCNMS, and we mosaicked the data at 1-meter pixel resolution. We integrated video from a towed camera sled, bathymetry data, sedimentary samples and side scan sonar mapping to describe geological and biological aspects of habitat. With a hierarchical deep-water marine benthic classification scheme, we created and attributed polygon features (Greene et al. 1999). This report provides a description of the mapping efforts and the results of the image classification procedure for each of the areas surveyed in 2004 and 2005.

Additionally, portions of these side scan sonar surveys partially overlapped a region of the sanctuary previously mapped with multibeam bathymetry from a survey which utilized a combination of Reson 8101 and 8125 echosounders as described in Intelmann et al. 2006. In that survey, radiometric and geometric corrections were applied to the multibeam backscatter (Beaudoin et al. 2002), and the side scan, ri theta and snippet packets were all processed applying across-track signal normalization to minimize the variations due to the angular response of the seafloor.

When considering the operational logistics of data acquisition in open coast environments that experience significant swell and chop, one can successfully acquire multibeam bathymetry at vessel speeds near 8 knots, in comparison to the typical 3.0 to 3.5 knots as normally targeted for acquiring traditional single beam side scan sonar imagery, such as with the model EG&G 272 or Klein 3000 which are both presently used for seabed mapping activities at OCNMS. Currently, with only an approximate 20 percent of the 8,200 $nm^2$ of the sanctuary adequately characterized, this difference in acquisition efficiency becomes an important consideration for meeting the ultimate goal of 100 percent sea floor characterization in a timely manner. Thus having a small sample of seafloor mapped with various methods provided an opportunity to assess classification results based on the two different types of acoustic imagery. With this, we hope to gain a better understanding of classification performance between the two types of imagery, helping to guide future mapping strategies at the OCNMS.

## SURVEY AREA

OCNMS conducted approximately 21 $km^2$ and 34 $km^2$ of seafloor mapping surveys aboard the *R/V TATOOSH* during the field seasons of 2004 and 2005, respectively. All survey areas were in the general vicinity of Cape Flattery, with exception of one southern block, located approximately 12 km offshore of Point of the Arches (Figure 1). We acquired survey lines from August 5 through 18 in 2004 and on May 16 and 17 and July 12 through 14 in 2005. Water depths ranged between 20 and 110 meters throughout the survey grounds.

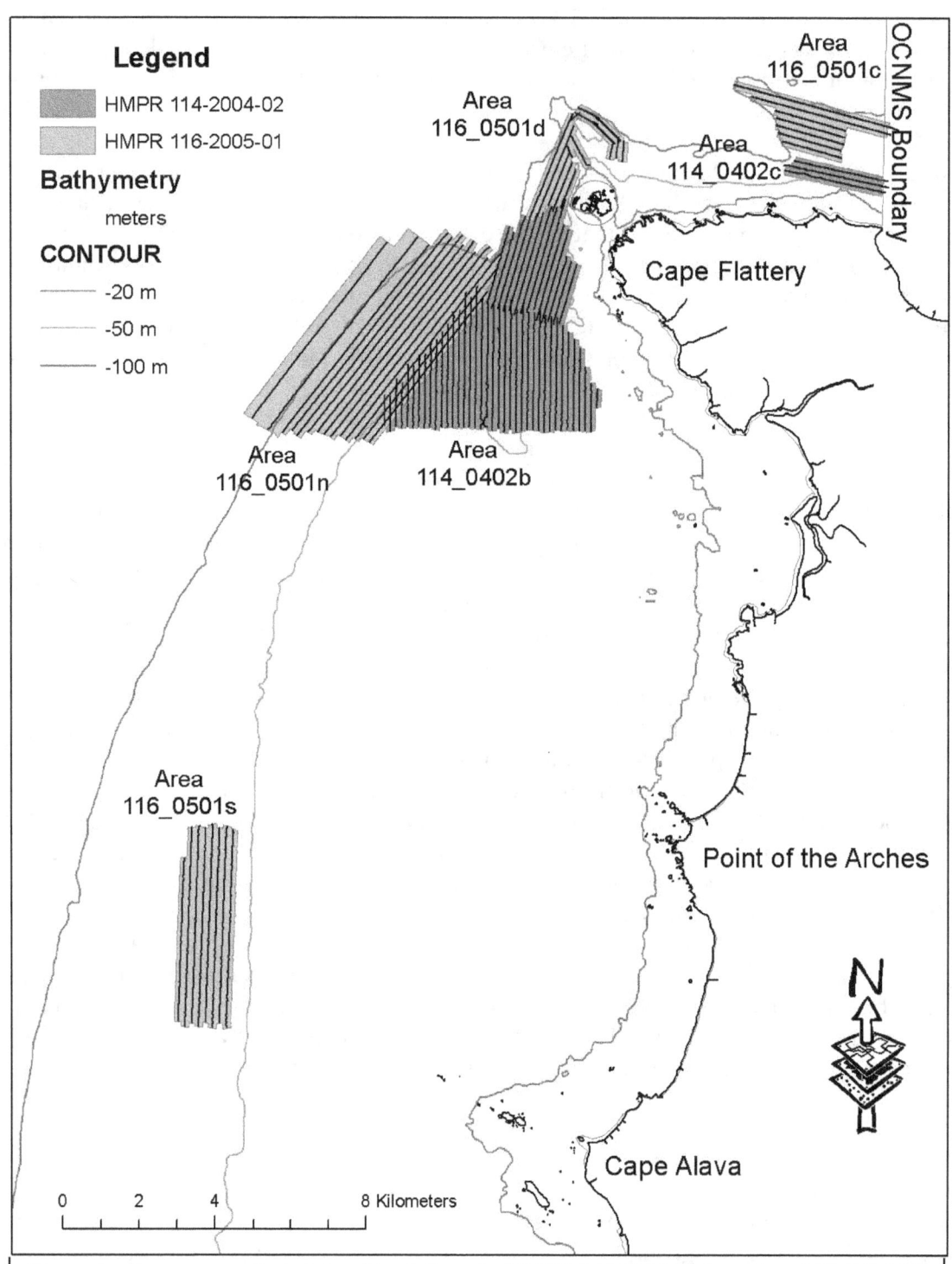

**Figure 1.** Sonar survey footprint and track lines for HMPR-114-2004-02 (green) and HMPR-116-2005-01 (orange) shown with selected isobaths.

## SONAR ACQUISITION AND DATA LOGGING

The NOAA research vessel *TATOOSH*, measuring 11.5 meters in length, served as the survey platform. We acquired ship positioning with a Trimble DSM 212L differential GPS (DGPS) and controlled line planning through Hypack Max software. We estimated towfish position through use of a digital cable counter, manufactured by Hydrographic Surveys, which logged line out.

**Figure 2**. Survey platform *R/V TATOOSH*.

We used an EG&G Model 272 analog side scan sonar to acquire the acoustic imagery (Figure 3). The sonar system has a horizontal beam width of 1.2 ° at 100 kHz with a vertical beam width measuring 50 °. We maintained vessel speed at between 3 and 3.5 knots throughout operations. We logged sonar imagery as 16 bit data with 1,024 samples per channel using Triton Imaging, Inc. (TII) Isis Sonar and recorded as eXtended Triton Format (XTF).

**Figure 3**. EG&G Model 272 towfish. Note magnetic cable counter attached to upper block housing.

We used an analog control interface (ACI) kit to convert the analog sonar signal to digital format plus provide individual-channel analog gain control of the towfish signal.

The 2004 survey was designed around a 150-meter line spacing plan, but in 2005 was increased to 175-meter spacing to reduce overlap and allow more ground to be covered over a given span of time. In both field seasons a 100-meter range scale was set on the sonar.

## SONAR DATA PROCESSING AND IMAGE CLASSIFICATION

The algorithm used to calculate towfish layback (horizontal distance behind the tow point) requires the input of an estimate of towfish depth (distance below water surface), towfish altitude (height above the sea floor) and the amount of cable out at any given time. Since this particular towfish was not equipped with a depth sensor, we did not precisely know the distance below the surface. As such, we had to estimate this value in some other way. To address this minor issue, we recorded water depth under keel as towfish depth in the XTF. After insuring proper bottom tracking, we used the Isis ASCII Report Utility to export a time stamp, depth (water depth) and towfish altitude from the XTF into a separate text file for each line of data. In an external spreadsheet, we simply subtracted towfish altitude from water depth to provide an estimate of continuous towfish depth. Although this method will not provide an absolutely accurate measurement for towfish depth, especially in areas of higher relief, the value will be relatively close because we designed the line planning to parallel the bathymetric contours to minimize differences in along track relief. Using the NavInXTF Utility, we imported the estimated towfish depth values back into the XTF files, thus replacing the previously logged water depth values. With final entry of the DGPS antennae offsets, we used a normal catenary layback calculation to provide an estimate of towfish position in the mosaics.

The navigation data was smoothed in Isis Sonar using a combination of a Kalman filter and a 7-point moving average filter. We accomplished slant range correction and bottom tracking in Isis Sonar, in addition to the application of time-varied gain and beam angle compensation curves.

We imported individual line mosaics into TII's DelphMap, merged them into separate mosaics for each survey block and then exported them as geotiff images. Image homogeneity and entropy were calculated for each mosaic using custom designed software (Cochrane and Lafferty 2002). Mosaics from the side scan packets, entropy and homogeneity images were all layer stacked in Erdas Imagine to create multi-spectral images. A supervised classification was performed using a maximum likelihood decision rule to produce a final classified image (Intelmann et al. (2006). Adobe Photoshop was then used to edit misclassified data such as that occurring nadir or in other various areas such as misclassified side lobes. Raster images were then smoothed with a low pass filter and converted to Features in ArcGIS.

## GROUNDTRUTHING

In August 2005, we used a custom designed camera sled (Figure 4) to acquire underwater videography for validating the sonar imagery (Intelmann et. al 2006). Although not available in every survey block, the usSeabed project (Reid et al. 2001) provided 34 samples as further weight of evidence for the video and sonar interpretation. Video transects and usSeabed sample locations are shown in Figure 5.

4

**Figure 4.** Towed camera sled used for groundtruthing efforts.

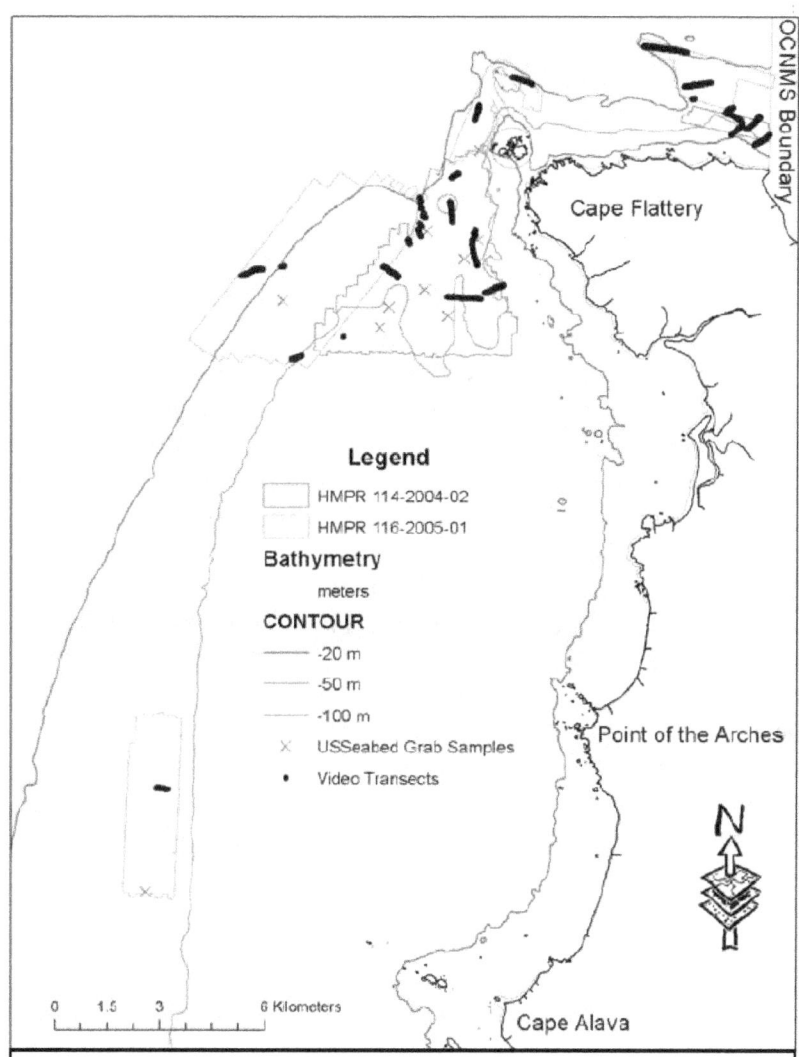

**Figure 5.** Location of video transects (black circles) and usSeabed sediment samples (blue cross-hair) shown with HMPR-114-2004-02 (green) and HMPR-116-2005-01 survey footprints and selected isobaths.

## SURVEY RESULTS AND INTERPRETATION

We acquired over 156 linear km of survey lines aboard the *R/V TATOOSH* in 2004 (Table 1), although we lost several potential days of survey time due to poor weather and/or various equipment challenges. We collected nearly 165 km of sonar survey lines in 2005, even though during the month of August, we spent the majority of the limited time available for habitat mapping related work aboard this same vessel conducting groundtruthing video surveys.

**Table 1.** Below are the survey effort statistics for HMPR-114-2004-02 and HMPR-116-2005-01. We acquired data aboard the *R/V TATOOSH* using an EG&G Model 272 side scan sonar. We surveyed two areas in 2004 and four areas in 2005. Area is presented in square kilometres, length of linear track lines in kilometers, and hours of actual logged sonar packets in hours, minutes, and seconds.

| Year | Block | Date | Area (km$^2$) | Tracks (km) | Hours (h:m:s) |
|------|-------|------|---------------|-------------|----------------|
| 2004 | 114_0402b | August 5-18 | 19.72 | 140.9 | 21:47:26 |
|      | 114_0402c | August 18 | 1.28 | 15.3 | 1:13:37 |
| **Total** | | | **21.00** | **156.2** | **23:01:03** |
| 2005 | 116_0501c | May 16-17 | 3.24 | 17.2 | 3:08:25 |
|      | 116_0501d | May 17 | 2.70 | 19.3 | 3:08:52 |
|      | 116_0501n | July 12-14 | 20.18 | 87.0 | 14:12:50 |
|      | 116_0501s | July 14 | 7.50 | 41.1 | 6:33:15 |
| **Total** | | | **33.62** | **164.6** | **27:03:22** |

We defined megahabitat categorization for all of the survey blocks as continental shelf (Greene et al. 1999). Survey block 116_0501s consisted entirely of soft (s), silty substrates (Table 2). Textural classification of the imagery suggested that mixed sediment (m) including cobbles, pebbles, gravel and boulders (mixed with soft substrate) characterized 80 percent of blocks 114_0402c and 116_0501c (both located offshore of Chibahdehl Rocks in the Strait of Juan de Fuca), while the remaining 20 percent of each of these two areas consisted of hard (h) complex rocky bottom (see Appendix for imagery). We classified block 116_0501d, which surrounded Duncan and Duntze Rocks and followed the western edge of Tatoosh Island, as 90 percent mixed substrate (see Appendix for imagery). Video imagery further revealed the surface to be mostly a combination of gravel, pebble, cobble and shell hash. The submerged basalt rock flanks of both Duncan and Duntze Rocks (Snavely et al. 1993) represent the remaining 10 percent of the habitat in this particular region.

Of the six areas surveyed, block 114_0402b contained the highest diversity of substrates. We classified six distinct outcrops, covering 18 percent of block 114_0402b, as hard complex rocky bottom. These hard areas are easily distinguishable in the multibeam bathymetry data as well as the side scan sonar imagery. Video observations confirmed these areas as being highly utilized by various species of rockfish and numerous other organisms (Figure 6). The imagery further reveals heavily tilted, and differentially eroded bedrock strands (Figure 7) resultant of anticlinal folding and thrust faulting occurring in the area (McCrory et al. 2004). Scattered areas of mixed substrate interspersed among bedrock strands define more than 37 percent of seafloor in this area.

In general, soft substrates occur in the southern portion of the survey block, and continue to the west throughout the majority of block 116_0501n (Figure 8). We easily delineated several areas of sediment waves, indicating active sediment movement occurring in specific areas.

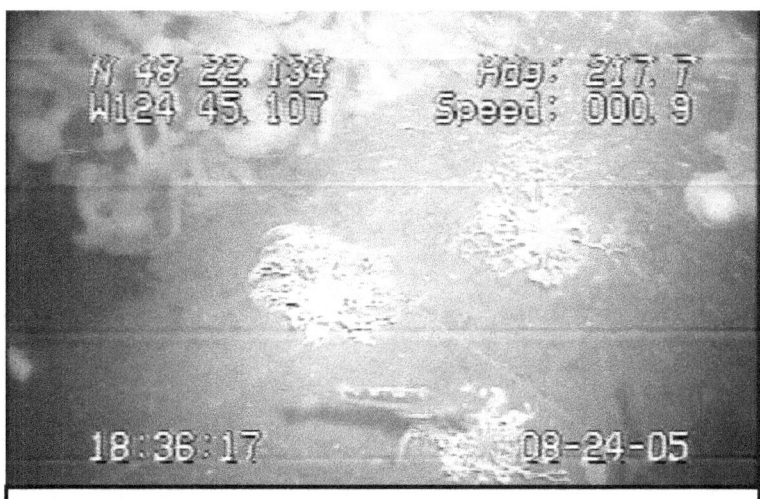

**Figure 6.** Example of hard, complex rocky bottom, providing a habitat for basket star, white-plumed anemone and rockfish.

**Table 2.** Distribution of bottom hardness for each sonar mosaic classified from survey HMPR-114-2004-02 and HMPR-116-2005-01. See Figure 1 for area locations. Bottom hardness codes are hard (h), mixed (m) and soft (s) – see previous section for description of classes. Area is presented in square meters (top value) and area as percentage of each individual mapped area (bottom bold value in the matrix).

| Year | Survey Block | h | m | s |
|---|---|---|---|---|
| 2004 | 114_0402b | 3,553,093.3 | 7,371,749.0 | 8,817,661.4 |
| | | **18.0** | **37.3** | **44.7** |
| | 114_0402c | 266,065.6 | 1,001,958.9 | 0.0 |
| | | **21.0** | **79.0** | **0.0** |
| 2005 | 116_0501c | 738,181.5 | 2,503,968.7 | 0.0 |
| | | **22.8** | **77.2** | **0.0** |
| | 116_0501d | 280,836.0 | 2,420,793.9 | 0.0 |
| | | **10.4** | **89.6** | **0.0** |
| | 116_0501n | 10,240.0 | 341,667.0 | 19,842,000.0 |
| | | **0.0** | **1.7** | **98.3** |
| | 116_0501s | 0.0 | 0.0 | 7,585,526.0 |
| | | **0.0** | **0.0** | **100.0** |

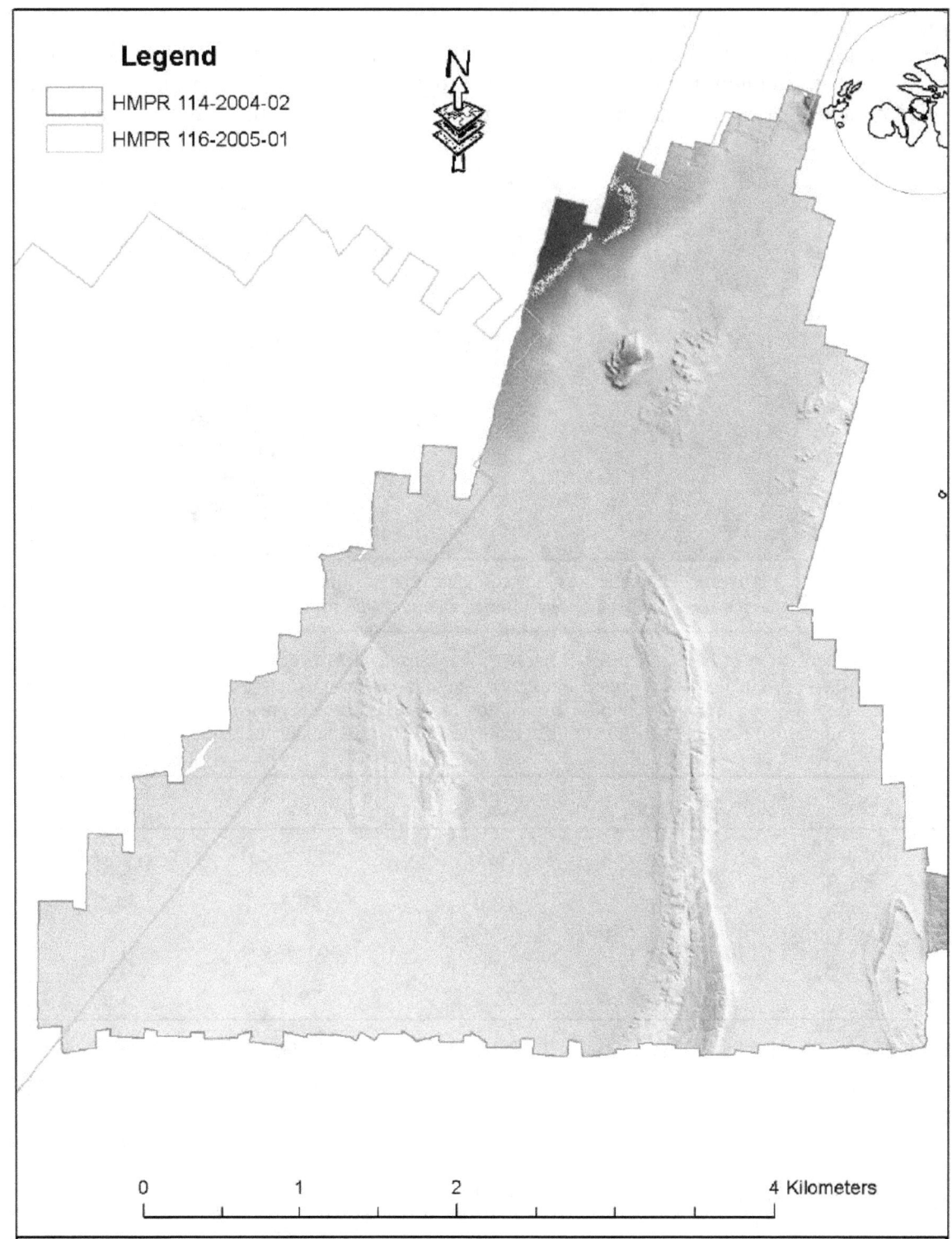

**Legend**

☐ HMPR 114-2004-02
☐ HMPR 116-2005-01

N

0    1    2              4 Kilometers

**Figure** 7. Digital terrain model showing multibeam bathymetry of survey block 114_0402b. Six distinct areas of high rugosity are easily distinguishable in the sun-illuminated bathymetry data. Note, however, that differences between soft and mixed substrates are not recognizable in this data. As such, side scan sonar imagery becomes the preferred data set for remotely delineating differences in these substrate classes.

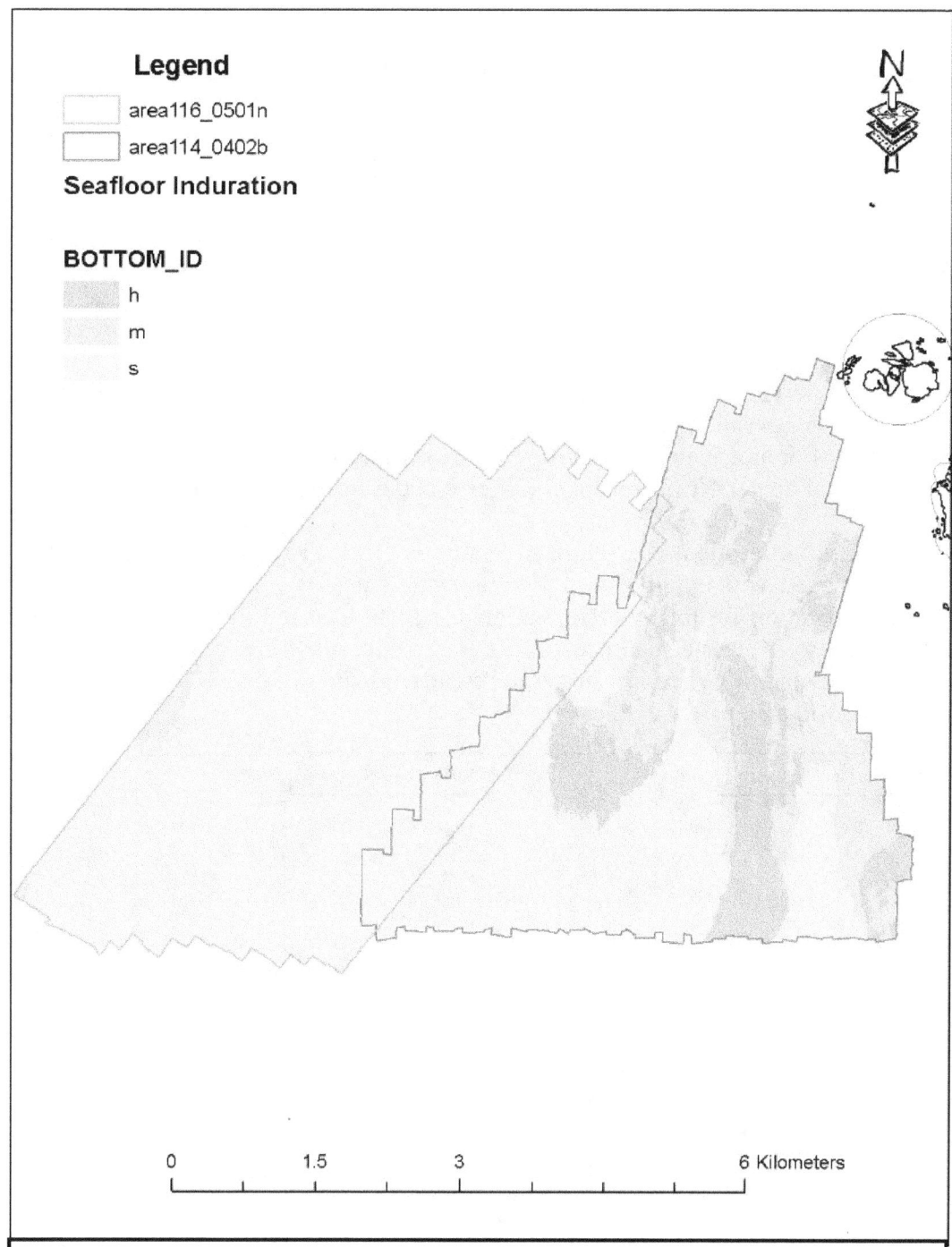

**Figure 8**. Bottom induration codes produced from textural classification of side scan sonar mosaics of survey blocks 114_0402b (green outline) and 116_0501n (orange outline). Note the ability to distinguish between mixed and soft substrates which was not possible through examination of the multibeam bathymetry data previously shown in Figure 7. h= hard substrate, m= mixed, and s= soft substrate.

The Appendices presents all side scan sonar mosaics, other attributed polygon layers showing bottom hardness (as in Figure 8) and matrix tables describing habitat classification for each of the six blocks surveyed in 2004 and 2005.

## DISCUSSION OF CLASSIFICATION BASED ON DIFFERENT SONAR TYPES

For all intents and purposes, side scan sonar and multibeam backscatter are essentially the same and are often spoken of interchangeably. Multibeam backscatter offers an advantage over traditional single beam side scan imagery in that the imagery is precisely geo-referenced, the systems are usually better calibrated and the data can be successfully acquired at much higher speeds when surveying in rough water. Multibeam echosounders, however, produce data based on much higher aspect ratios, and as such generally yield poorer resolution imagery than is possible through traditional single beam side scan sonar methods. Therefore, important image textural properties (including shadows) are often lost with hull-mounted multibeam echosounder systems, making for increased challenges during the classification procedure. This is especially noteworthy because textural homogeneity and entropy are two key components used by OCNMS in the classification process (Cochrane and Lafferty 2002; Intelmann and Cochrane 2006).

Of the 19.74 km$^2$ of seafloor characterized in survey block 114_0402b, 4.5 km$^2$ were also previously imaged with multibeam echosounder (Figure 9). In comparing results of textural classification for this overlapping area, which occurred in depths to 50 meters, overall results were nearly identical (Table 3). Only small areas of mixed sediment within the rock outcrops were lost in the backscatter method of classification but general features were still adequately delineated.

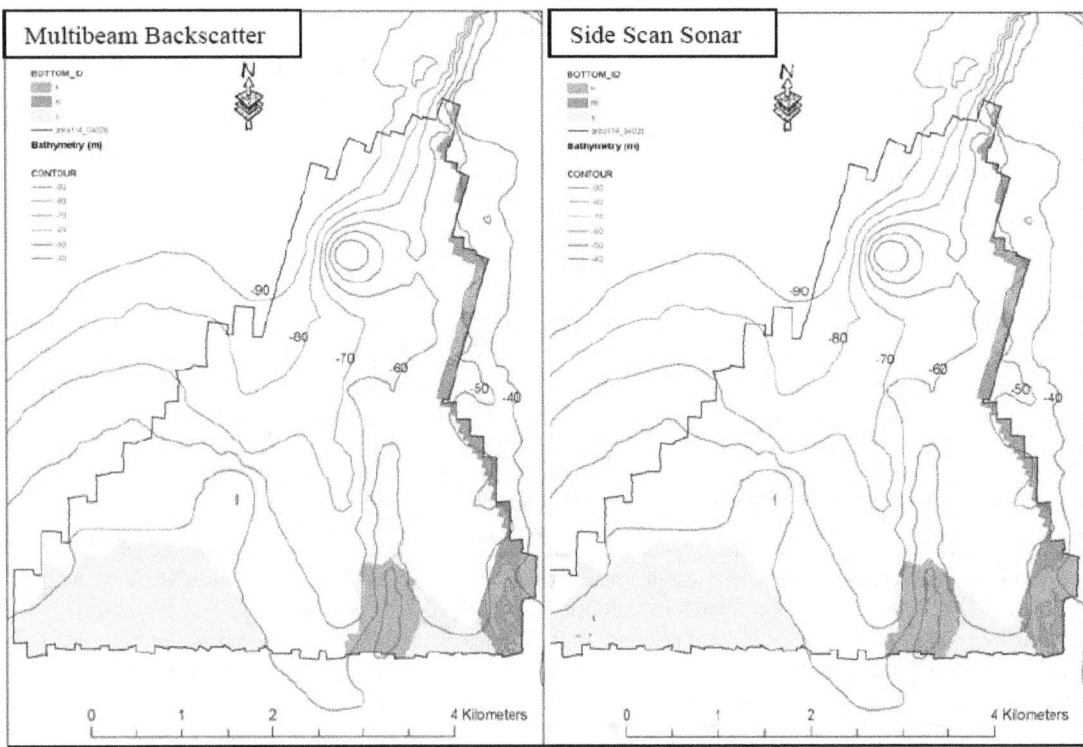

**Figure 9.** This figure is a footprint of survey block 114_0402b (black outline) showing overlapping areas mapped with both types of sonar (shaded polygons). Results of texture classification are based on side scan sonar (right) and multibeam backscatter (left). Note the overall similar results with only small areas of mixed sediment recognizable in the side scan imagery being lost in the backscatter classification. Tan=soft (s), green = mixed (m), and red = hard (h).

**Table 3.** This table is a comparison of classification results of bottom hardness for the overlapping area within survey block 114_0402b mapped with both side scan sonar and multibeam echosounder. Bottom hardness codes are hard (h) and mixed (m), and soft (s). Area is presented in square meters (top value) and area as percentage of each individual mapped area (bottom bold value in the matrix).

| Substrate Class | Side Scan Sonar | Multibeam Backscatter |
|---|---|---|
| h | 257,717.99 | 269,652.06 |
| | **16.23** | **16.99** |
| m | 1,330,012.99 | 1317856.23 |
| | **83.77** | **83.01** |
| m | 1,330,012.99 | 1317856.23 |
| | **83.77** | **83.01** |

As with the area of 114_0402b, the classification results for two methods of data acquisition that overlapped in survey block 116_0501d compared well (Figure 10). Of the 2.7 km$^2$ of seafloor characterized by side scan sonar in survey block 116_0501d, 1.58 km$^2$ were also previously imaged with multibeam echosounder (Table 4). Results of this small dataset are comparable to 100-meters depth in this region.

**Figure 10.** Footprint of survey block 116_0501d (black outline) showing overlapping areas mapped with both types of sonar (shaded polygons). Results of texture classification are based on side scan sonar imagery (left) and multibeam backscatter (right). Note the overall similar results with only one small area of hard bottom recognizable in the side scan imagery being lost in the backscatter classification. green = mixed (m), and red = hard (h).

11

**Table 4.** This table is a comparison of classification results of bottom hardness for the overlapping area within survey block 116_0501d mapped with both side scan sonar and multibeam echosounder. Bottom hardness codes are hard (h) and mixed (m). Area is presented in square meters (top value) and area as percentage of each individual mapped area (bottom bold value in the matrix).

| Substrate Class | Side Scan Sonar | Multibeam Backscatter |
|---|---|---|
| h | 257,717.99 | 269,652.06 |
| | **16.23** | **16.99** |
| m | 1,330,012.99 | 1,317,856.23 |
| | **86.77** | **83.01** |

More noticeable differences became apparent when examining the classification results between the two methods within survey block 114_0402c (Table 5). Of the 1.26 km$^2$ of seafloor characterized by side scan sonar in survey block 114_0402c, 100 percent of this same area was previously surveyed with multibeam echosounder. Although the general shape of the major rock feature in this area remained similar (Figure 11), when compared to classification results based on backscatter from multibeam echosounder data, side scan sonar classification produced an increase in hard bottom substrate of nearly 7 percent.

**Table 5.** This table is a comparison of classification results of bottom hardness for the overlapping area within survey block 114_0402c mapped with both side scan sonar and multibeam bathymetry. Bottom hardness codes are hard (h) and mixed (m). Area is presented in square meters (top value) and area as percentage of each individual mapped area (bottom bold value in the matrix).

| Substrate Class | Side Scan Sonar | Multibeam Backscatter |
|---|---|---|
| h | 266,065.56 | 179,146.10 |
| | **20.98** | **13.99** |
| m | 1,001,58.88 | 1,101,063.00 |
| | **79.02** | **86.01** |

Beyond the 30-meter isobath, the textural properties of the mutlibeam backscatter imagery were far inferior to the imagery produced by the side scan sonar (Figure 12). A combination of the seafloor slope, a thin superficial layer of fine sediment on the rock surface as evident from video imagery and the high aspect ratio associated with the multibeam bathymetry (in comparison to towed side scan sonar which placed the sonar closer to the seafloor), are all likely causes for this reduction of image textural enhancement. Although the image classification performance of the two different sonar methods were comparable in various regions of survey blocks 114_0402b and 116_0501d, the loss of image resolution in the multibeam backscatter data of block 114_0402c is the main drawback to using this type of data to create habitat maps in OCNMS.

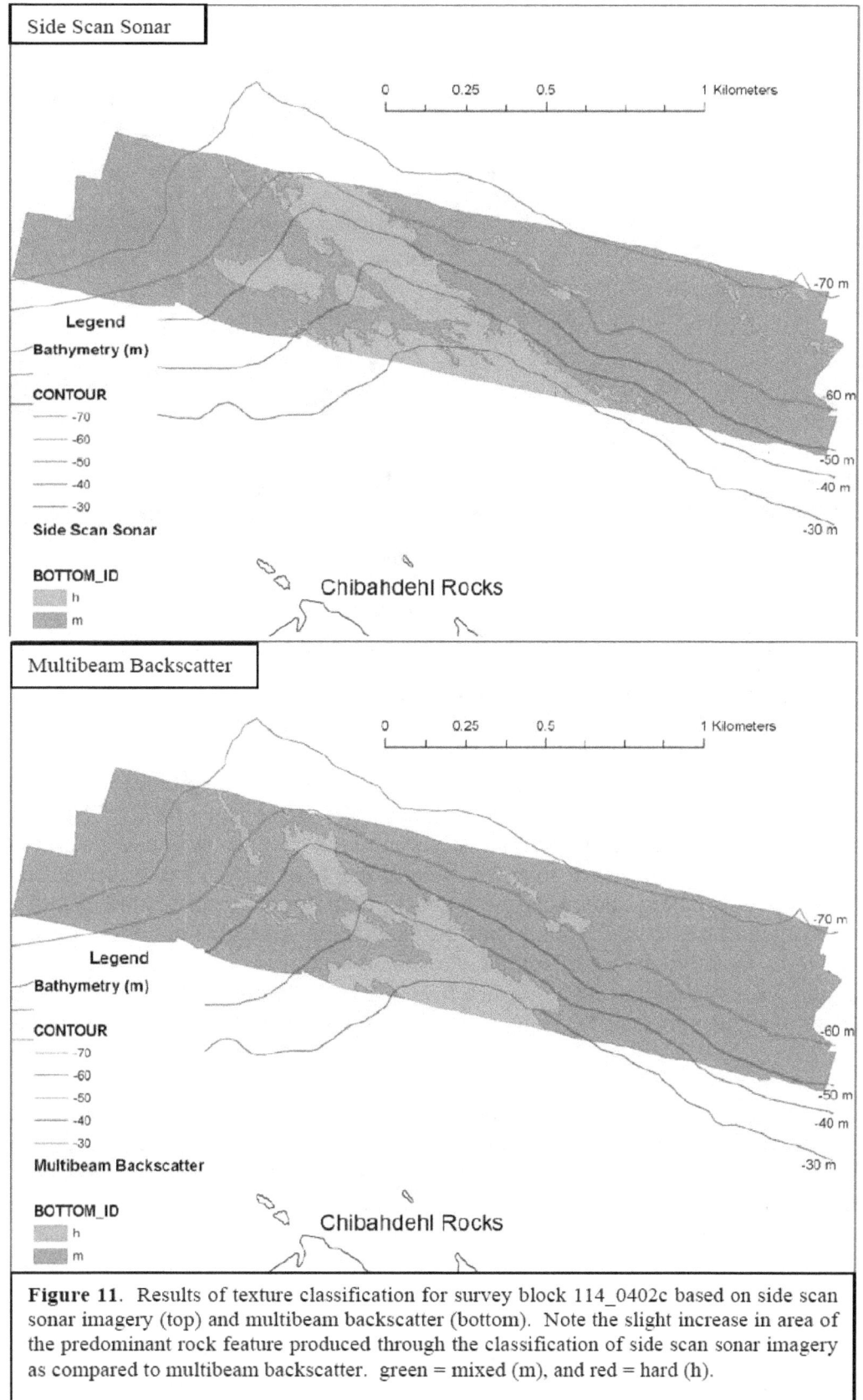

**Figure 11.** Results of texture classification for survey block 114_0402c based on side scan sonar imagery (top) and multibeam backscatter (bottom). Note the slight increase in area of the predominant rock feature produced through the classification of side scan sonar imagery as compared to multibeam backscatter. green = mixed (m), and red = hard (h).

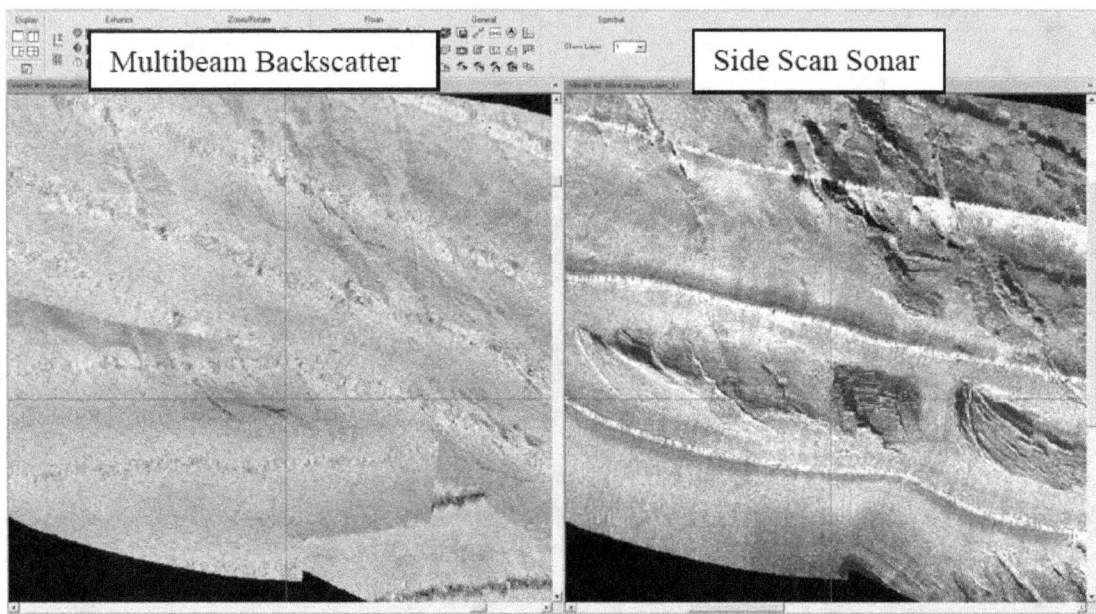

**Figure 12.** Multibeam backscatter imagery (left) compared with side scan sonar imagery (right) for the same area within survey block 114_0402c. The enhanced textural properties associated with the side scan sonar imagery are easily identifiable. The images are geospatially linked with the red cross hair which represents the same geographic position in each image.

Unlike side scan sonar, when using multibeam echosounder systems swath width will decrease with decreasing water depths (Kashomita et al. 2005). This reduced area of coverage, however, can be offset by the increased speed of acquisition when comparing to use of a traditional single beam side scan sonar. Moreover, the ubiquitous unexposed rock pinnacles and outcrops occurring within the 30-meter isobath in OCNMS create dangers to towed side scan sonar gear that are not realized with multibeam echosounder systems. In depths shallower than 30-meters, the high aspect ratio associated with multibeam bathymetry systems appears to not degrade results of the texture classification procedure being used at OCNMS. But, the degrading effects of this high aspect ratio associated with multibeam echosounders would become even more pronounced with increasing water depths since the distance from the sonar to the seafloor would be greatly increased in a hull-mounted technique as compared to a towed side scan sonar scenario.

Due to the swath width/depth dependency associated with multibeam bathymetry systems, it makes even more sense to use multibeam backscatter for surveying in deeper water, only to return with more labor intensive deep-towed side scan sonar in areas where the poorer resolution hull-mounted multibeam bathymetry systems suggest features of interest exist. In deep water (> 200 meters), traditional single beam side scan sonar imagery would be best considered as a complement to multibeam bathymetry (and backscatter) to further interrogate areas of extreme interest that require increased resolution.

It is important to note that more recent technology has led to the production of multi-beam side scan sonar (such as the Klein 5000 series) which function through use of electronic phase delay to accomplish beam steering and allows for successful data acquisition to speeds approaching 10 knots. NOAA's Hydrographic Survey Division has even successfully hull-mounted these newer generation sonar models on survey launches and presently has several of these specific configurations in operation throughout the fleet. Use of this newer generation multi-beam side scan sonar could provide the best of both worlds in shallow water due to the high resolution and existing potential for high speed acquisition. But there are also several drawbacks to using multi-beam side scan sonar that warrant consideration, namely cost (a Klein 5000, for example, is nearly four times as costly as a Klein 3000) and size. Unlike deployment of a traditional single beam side scan sonar, the multi-beam side scan sonar systems are much larger in size and generally require multiple individuals to handle which make them far more unwieldy to deploy and retrieve in a towable configuration off a small vessel.

As previously mentioned, these newer generation systems can be hull-mounted but they then lose utility in water depths much greater than 30 meters since they are only currently available to the commercial market as 455 kHz systems, and as such suffer from signal attenuation when used in deeper waters as a hull-mount. Furthermore, unlike towed side scan sonar a hull-mounted side scan sonar would also be subject to greater geometric distortions created by excessive vessel movement which would otherwise be compensated through attitude corrections in multibeam bathymetry. Since the open coastal environment within OCNMS is often plagued with annoying wind chop and confused seas, it is almost certain that vessel movement would propagate into the imagery when used in a hull-mounted configuration in this area. As a towed setup much of this distortion would be removed since the tow cable would absorb much of the vessel pitch and roll.

With these considerations, backscatter derived from multibeam bathymetry is currently a cost-efficient and safe method for seabed imaging in the shallow (<30 meters) rocky waters of OCNMS. The image quality is sufficient for classification purposes, the associated depths provide further descriptive value and risks to gear are minimized. In shallow waters (<30 meters) which do not have a high incidence of dangerous rock pinnacles, a towed multi-beam side scan sonar could provide a better option for obtaining seafloor imagery due to the high rate of acquisition speed and high image quality. A hull-mounted multi-beam side scan sonar would likely suffer from image degradation due to vessel movement in this environment and is not an option in deeper water as the range scale is limited to 150 meters. Additionally, the high probability of losing or damaging such a costly system when deployed as a towed configuration in the extremely rugose nearshore zones within OCNMS is a financially risky proposition.

The development of newer technologies such as intereferometric multibeam systems and bathymetric side scan systems could also provide great potential for mapping these nearshore rocky areas as they allow for high speed data acquisition, produce precisely geo-referenced side scan imagery to bathymetry, and do not experience the angular depth dependency associated with multibeam echosounders allowing larger range scales to be used in shallower water. As such, further investigation of these systems is needed to assess their efficiency and utility in these environments compared to traditional side scan sonar and multibeam bathymetry.

## ACKNOWLEDGMENTS

The authors wish to thank Dave Kirner, Mike Levine, and Andy Palmer for assisting with survey mobilization, deployment and retrieval of the towfish and camera sled during survey operations, and for operating the R/V TATOOSH.

## REFERENCES

Beaudoin, J., Hughes Clarke, J.E., Van den Ameele, E. and Gardner, J., 2002, Geometric and radiometric correction of multibeam backscatter derived from Reson 8101 systems: Canadian Hydrographic Conference 2002 Proceedings (CDROM), Toronto, Canada.

Cochrane, G.R., and K.D. Lafferty. 2002. Use of acoustic classification of sidescan sonar data for mapping benthic habitat in the Northern Channel Islands, California. Continental Shelf Research 22: 683-690.

Greene, H.G., M.M. Yoklavich, R.M. Starr, V.M. O'Connell, W.W. Wakefield, D.E. Sullivan, J.E. McRea, Jr., G.M. Cailliet. 1999. A classification scheme for deep seafloor habitats. Oceanologica Acta. 22(6):663

Intelmann, S.S. and G.R. Cochrane. 2005. Benthic Habitat Mapping in the Olympic Coast National Marine Sanctuary: Classification of side scan sonar data from survey HMPR-108-2002-01: Version I. Marine Sanctuaries Conservation Series MSD-05-07. U.S. Department of Commerce, National Oceanic and Atmospheric Administration, Marine Sanctuaries Division, Silver Spring, MD. 13pp.

Intelmann, S.S., J. Beaudoin, and G.R. Cochrane. 2006. Normalization and characterization of multibeam backscatter: Koitlah Point to Point of the Arches, Olympic Coast National Marine Sanctuary - Survey HMPR-115-2004-03. Marine Sanctuaries Conservation Series MSD-06-03. U.S. Department of Commerce, National Oceanic and Atmospheric Administration, Marine Sanctuaries Division, Silver Spring, MD. 22pp.

Intelmann, S.S. 2006. Comments on Hydrographic and Topographic LIDAR Acquisition and Merging with Multibeam Sounding Data Acquired in the Olympic Coast National Marine Sanctuary. Marine Sanctuaries Conservation Series MSD-06-05. U.S. Department of Commerce, National Oceanic and Atmospheric Administration, Marine Sanctuaries Division, Silver Spring, MD. 17pp.

Kamoshita, T., Y. Sato, and T. Komatsu. 2005. Hydro-Acoustic Survey Scheme for Sea-Bottom Ecology Mapping. Seatechnology 46(6): 39-43.

McCrory, P.A., S.C. Wolf, S.S. Intelmann, W.W. Danforth, R.J. Weldon, J.L. Blair. 2004. Quaternary tectonism in a collision zone, Northwest Washington. Eos Trans. AGU, 85(47), Fall Meet. Suppl., Abstract T33C-1391.

Reid, J. A., Jenkins, C., Field, M. E., Gardner, J. V. and Box, C. E. 2001. USSEABED: defining the surface geology of the continental shelf through data integration and fuzzy set theory. Geological Society of America Annual Meeting, Boston, MA. Abstracts with Programs 33:A106.

Snavely, P.D., Jr., N.S. MacLeod, and A.R. Niem. 1993. Geologic Map of the Cape Flattery, Clallam Bay, Ozette Lake, and Lake Pleasant Quadrangles, Northwestern Olympic Peninsula, Washington. Map I-1946. U.S. Geological Survey.

# APPENDIX

## Appendix 1.  Isis Processing Parameters

<u>HMPR-114-2004-02</u>
Lateral Offset:  0.0m
Layback Offset:  6.7m
Heading= use CMG
Mosaic resolution: 0.3m (later reduced to 1m)
Apply BAC
TVG: start at first return
    Curve = -4 +0.75 + (-2)

<u>HMPR-116-2005-01</u>
***area116_0501c*
Lateral Offset:  0.0m
Layback Offset:  6.7m
Heading= use CMG
Mosaic resolution: 0.3m (later reduced to 1m)
Apply BAC
TVG: start at first return
    Curve = -7 + 0.09 + (-1)

***area116_0501d*
Lateral Offset:  0.0m
Layback Offset:  6.7m
Heading= use CMG
Mosaic resolution: 0.3m (later reduced to 1m)
Apply BAC
TVG: start at first return
    Curve = -5 + 0.08 + (0)

***area116_0501n and area116_0501s*
Lateral Offset:  0.0m
Layback Offset:  6.7m
Heading= use CMG
Mosaic resolution: 0.5m (later reduced to 1m)
Apply BAC
TVG: start at first return
    Curve = -5 + 0.09 + (1)

**Appendix 2.  Side Scan Sonar Imagery**

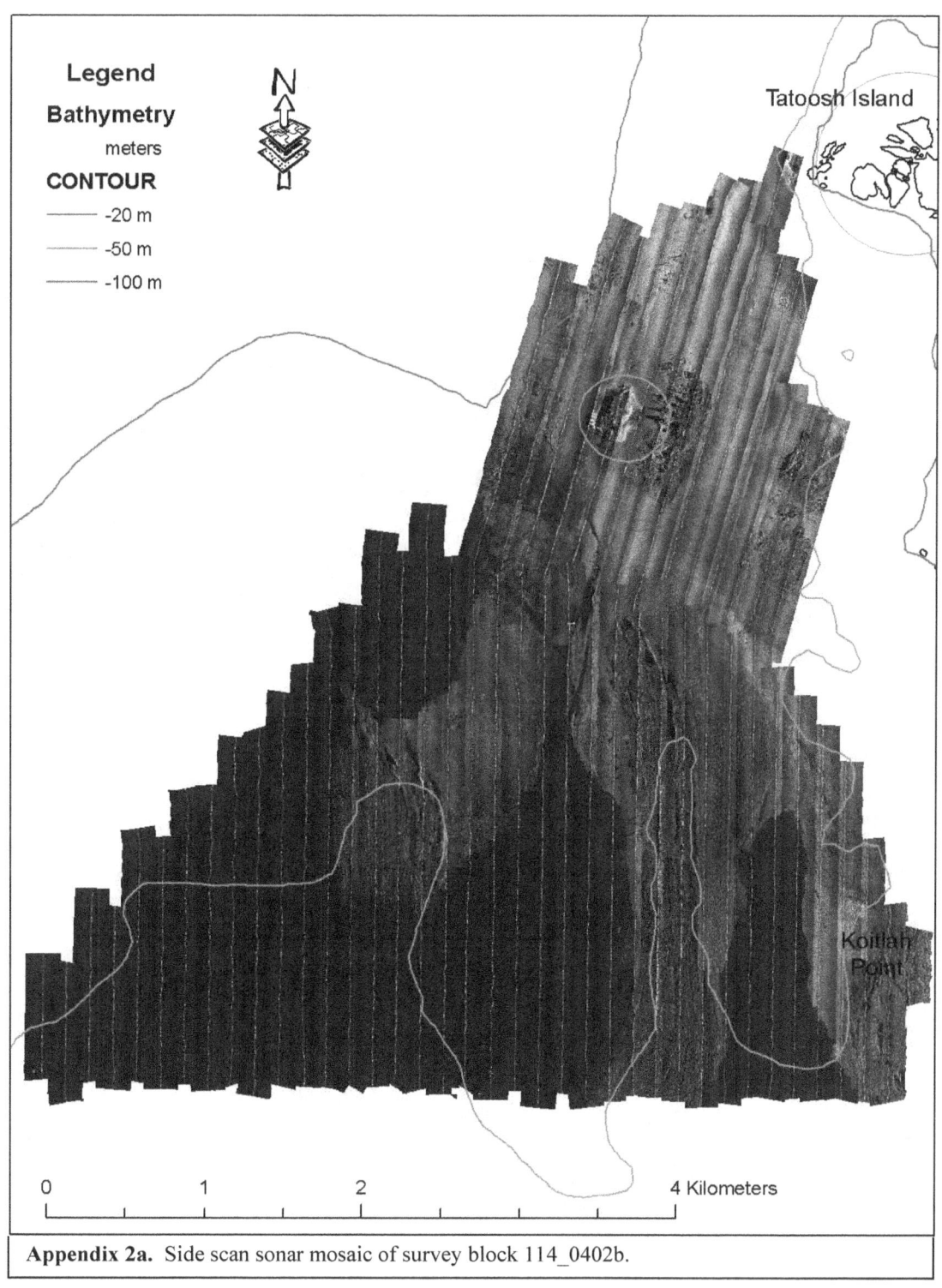

**Appendix 2a.** Side scan sonar mosaic of survey block 114_0402b.

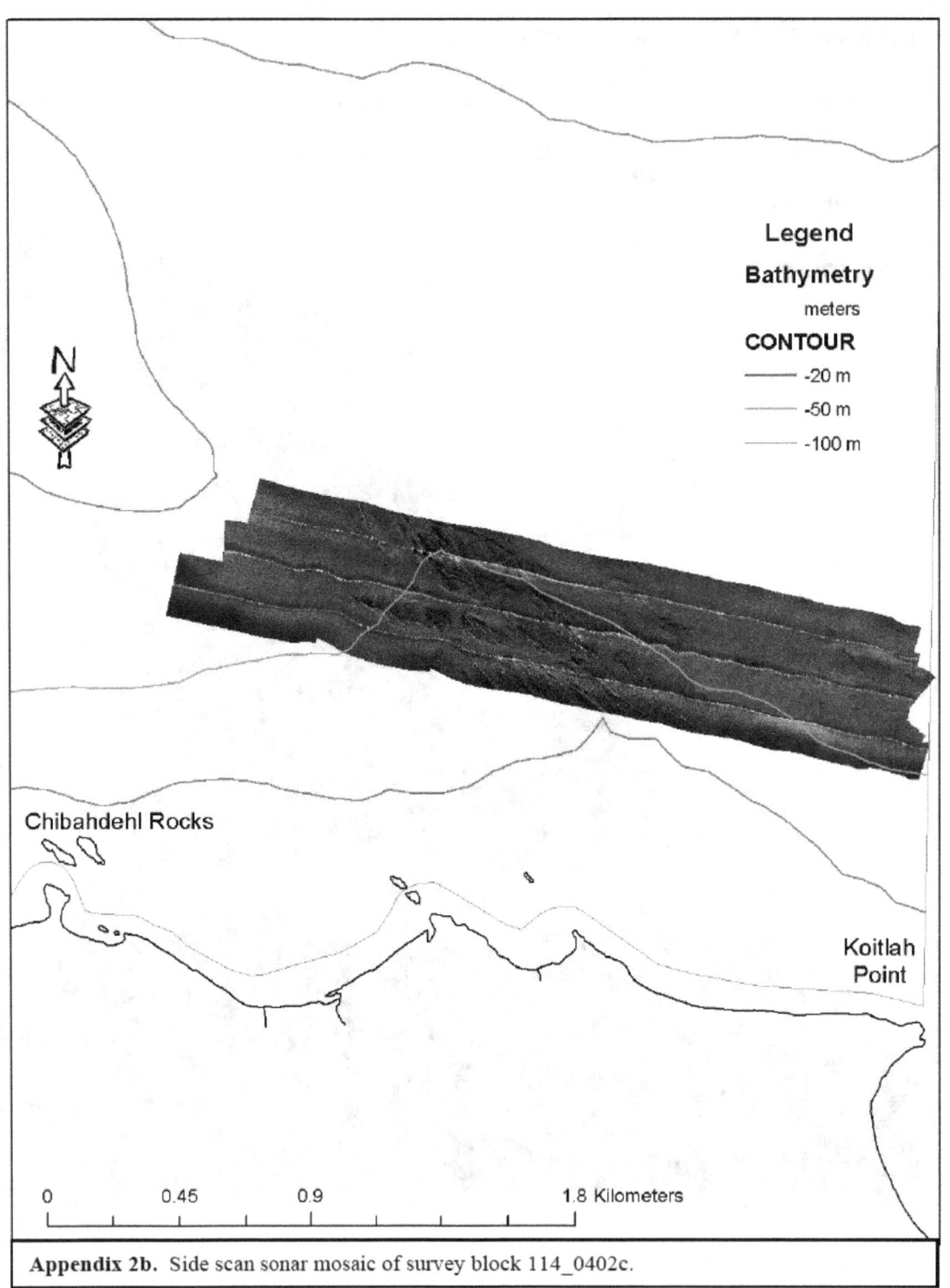

Legend

**Bathymetry**

meters

**CONTOUR**

——— -20 m

——— -50 m

——— -100 m

N

Chibahdehl Rocks

Koitlah
Point

| 0 | 0.45 | 0.9 | | 1.8 Kilometers |

**Appendix 2b.** Side scan sonar mosaic of survey block 114_0402c.

20

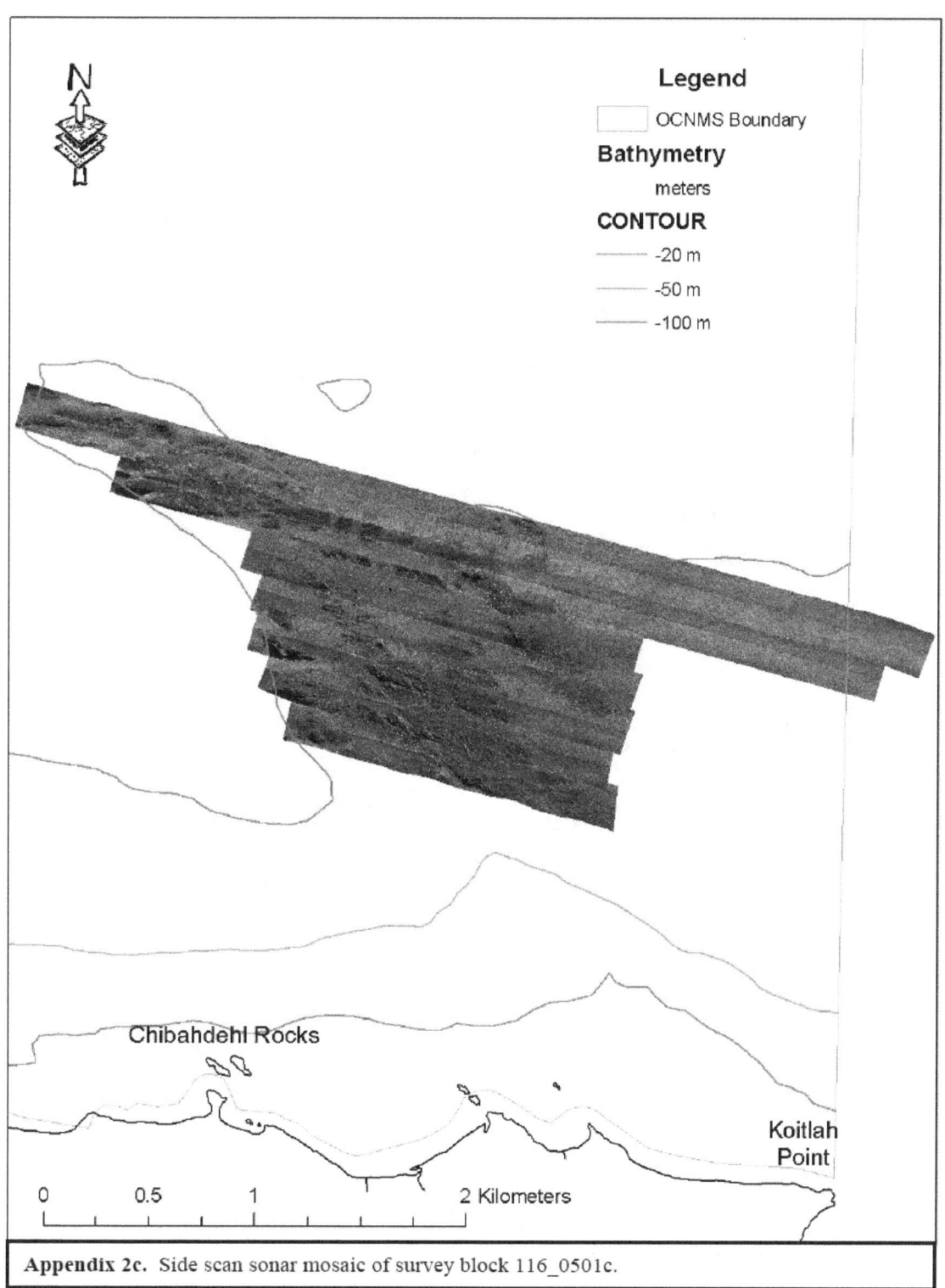

**Legend**

□ OCNMS Boundary

**Bathymetry**

meters

**CONTOUR**

—— -20 m

—— -50 m

—— -100 m

Chibahdehl Rocks

Koitlah Point

| 0 | 0.5 | 1 | 2 Kilometers |

**Appendix 2c.** Side scan sonar mosaic of survey block 116_0501c.

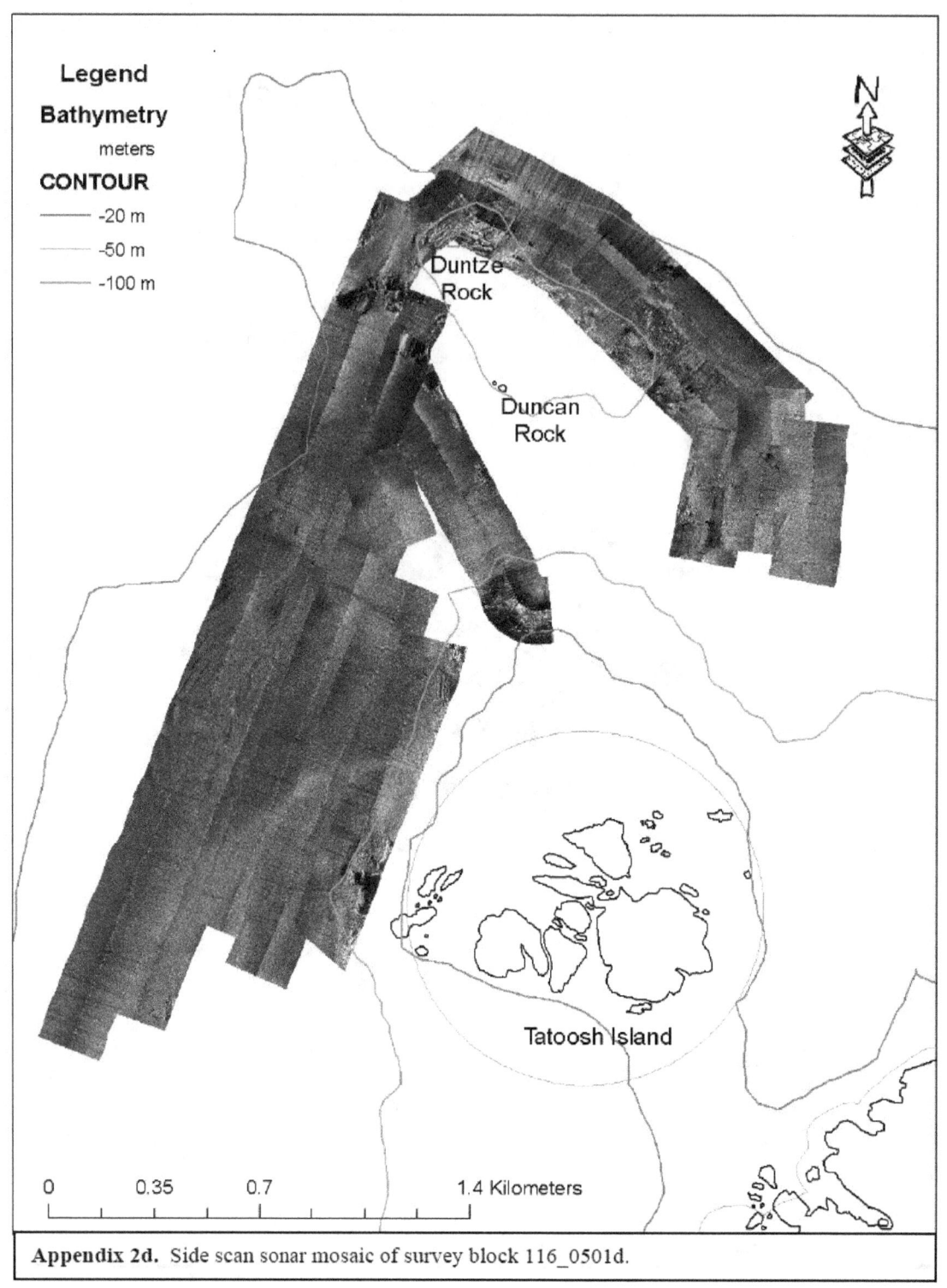

**Legend**

**Bathymetry**

meters

**CONTOUR**

——— -20 m

——— -50 m

——— -100 m

Duntze Rock

Duncan Rock

Tatoosh Island

0    0.35    0.7    1.4 Kilometers

**Appendix 2d.** Side scan sonar mosaic of survey block 116_0501d.

**Legend**

**Bathymetry**

meters

**CONTOUR**

——— -20 m

——— -50 m

——— -100 m

N

| 0 | 1 | 2 | | 4 Kilometers |

**Appendix 2e.** Side scan sonar mosaic of survey block 116_0501n.

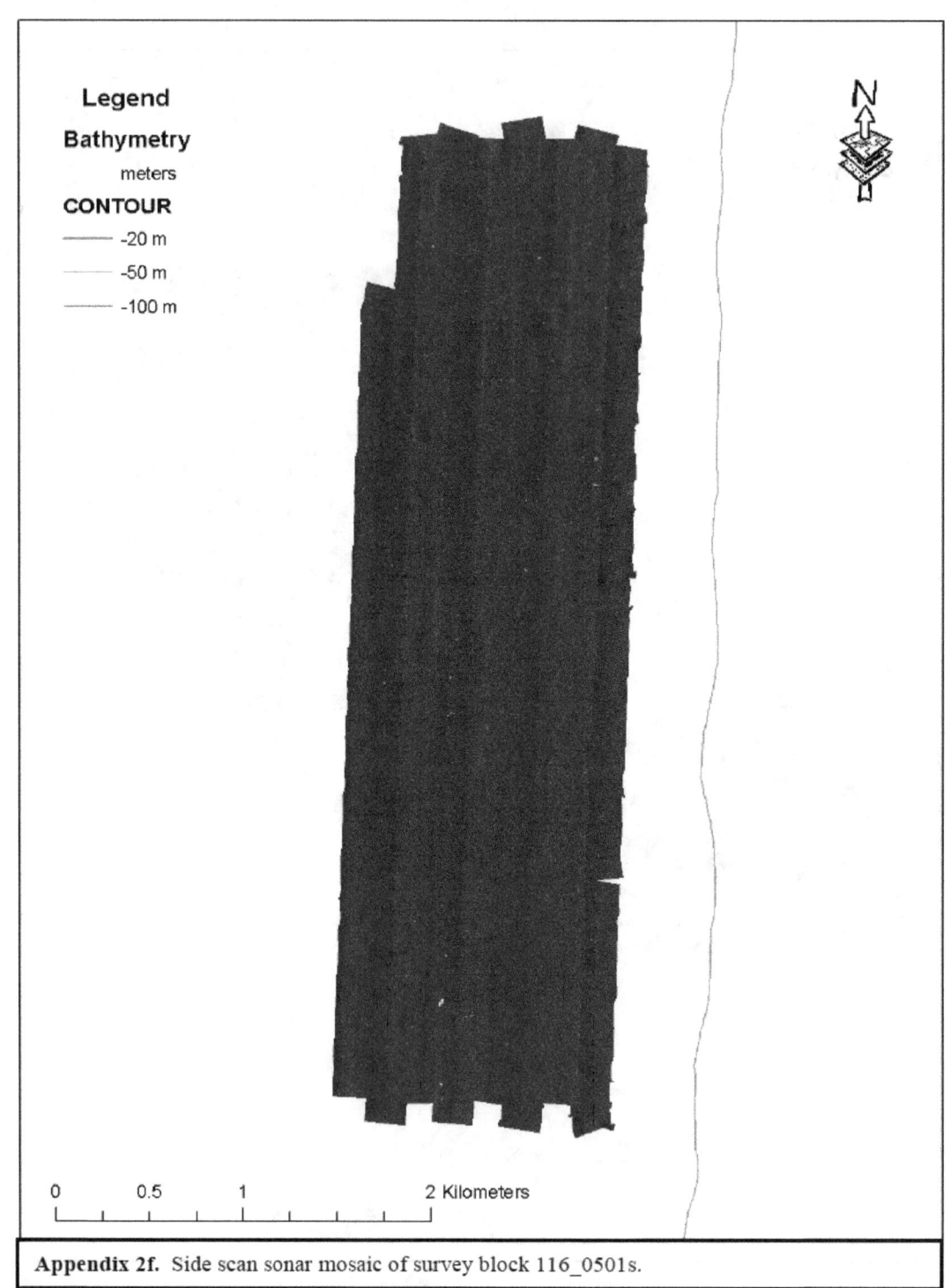

**Appendix 2f.** Side scan sonar mosaic of survey block 116_0501s.

## Appendix 3. Habitat Classification Polygons

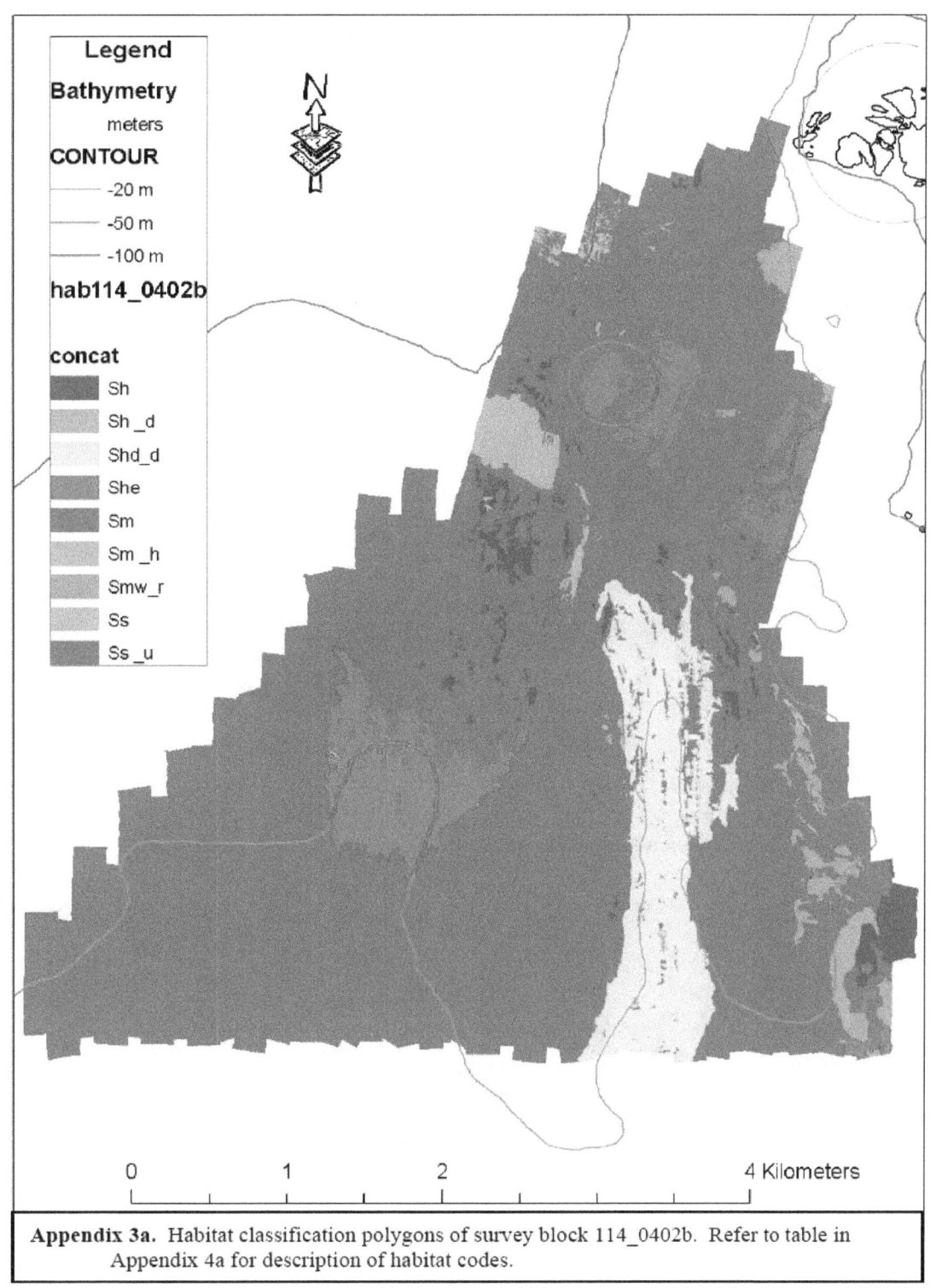

**Appendix 3a.** Habitat classification polygons of survey block 114_0402b. Refer to table in Appendix 4a for description of habitat codes.

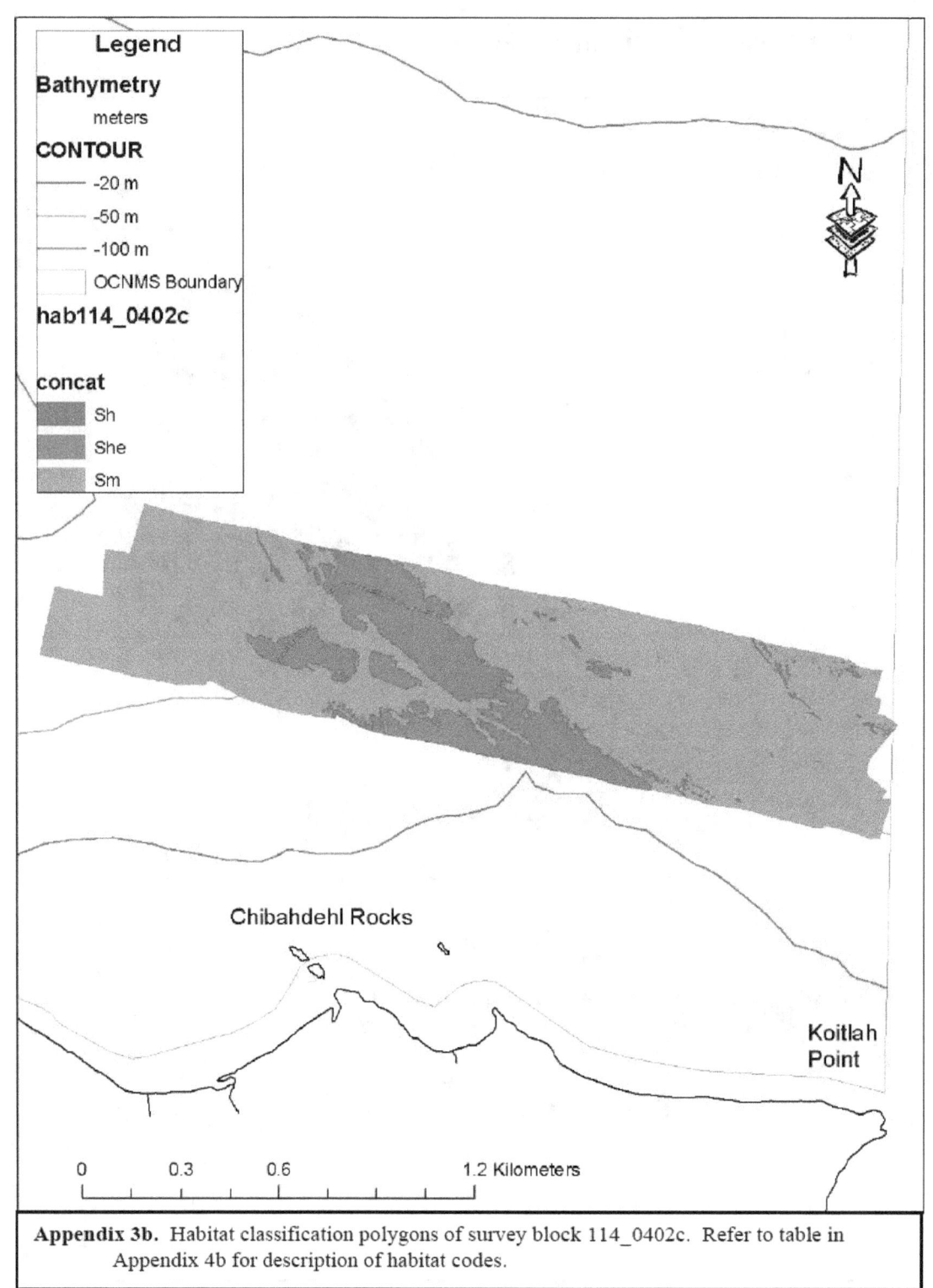

**Appendix 3b.** Habitat classification polygons of survey block 114_0402c. Refer to table in Appendix 4b for description of habitat codes.

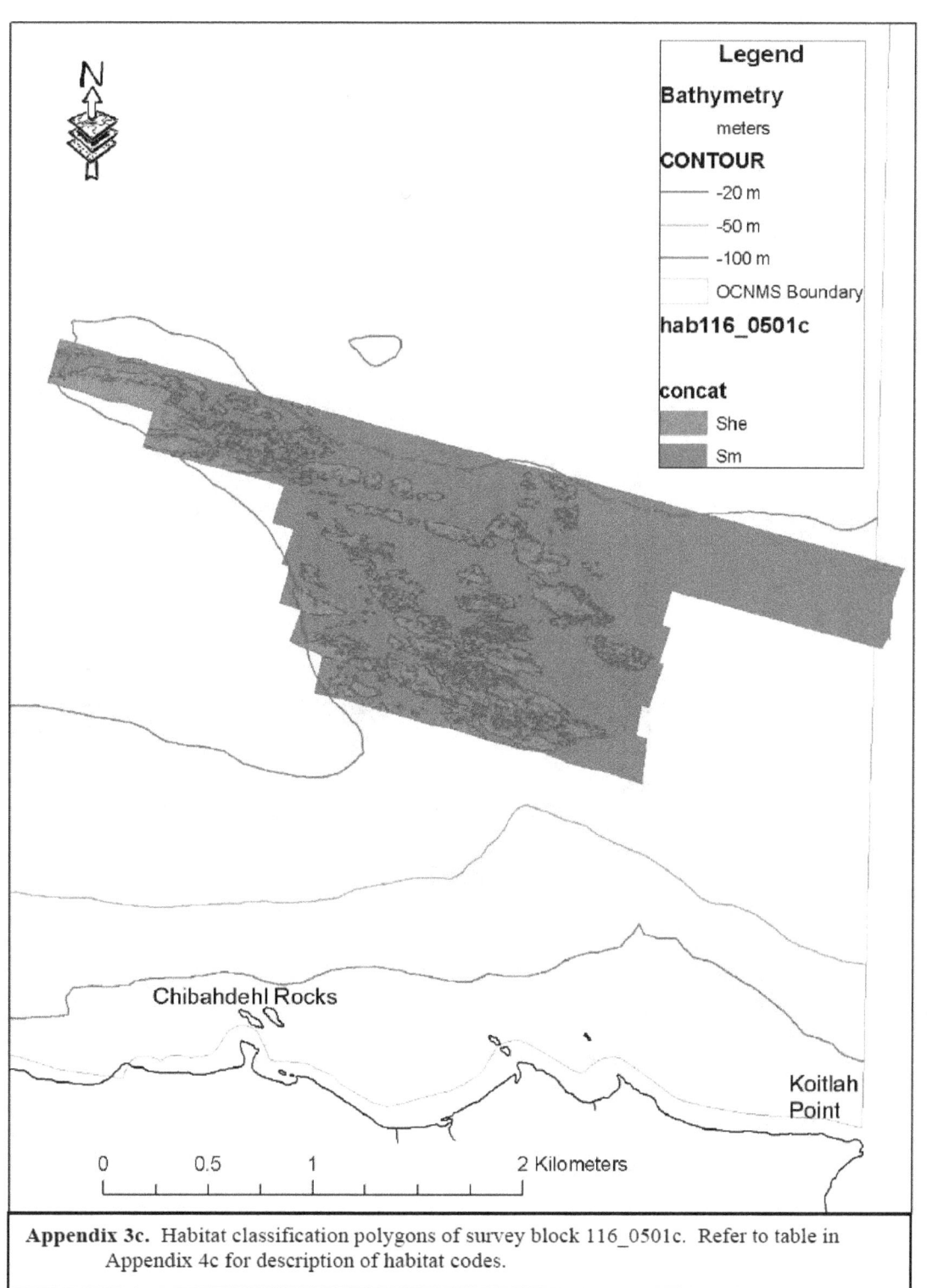

**Legend**

**Bathymetry**
meters
**CONTOUR**
—— -20 m
—— -50 m
—— -100 m
☐ OCNMS Boundary

**hab116_0501c**

**concat**
☐ She
☐ Sm

Chibahdehl Rocks

Koitlah
Point

0      0.5      1                2 Kilometers

**Appendix 3c.** Habitat classification polygons of survey block 116_0501c. Refer to table in
Appendix 4c for description of habitat codes.

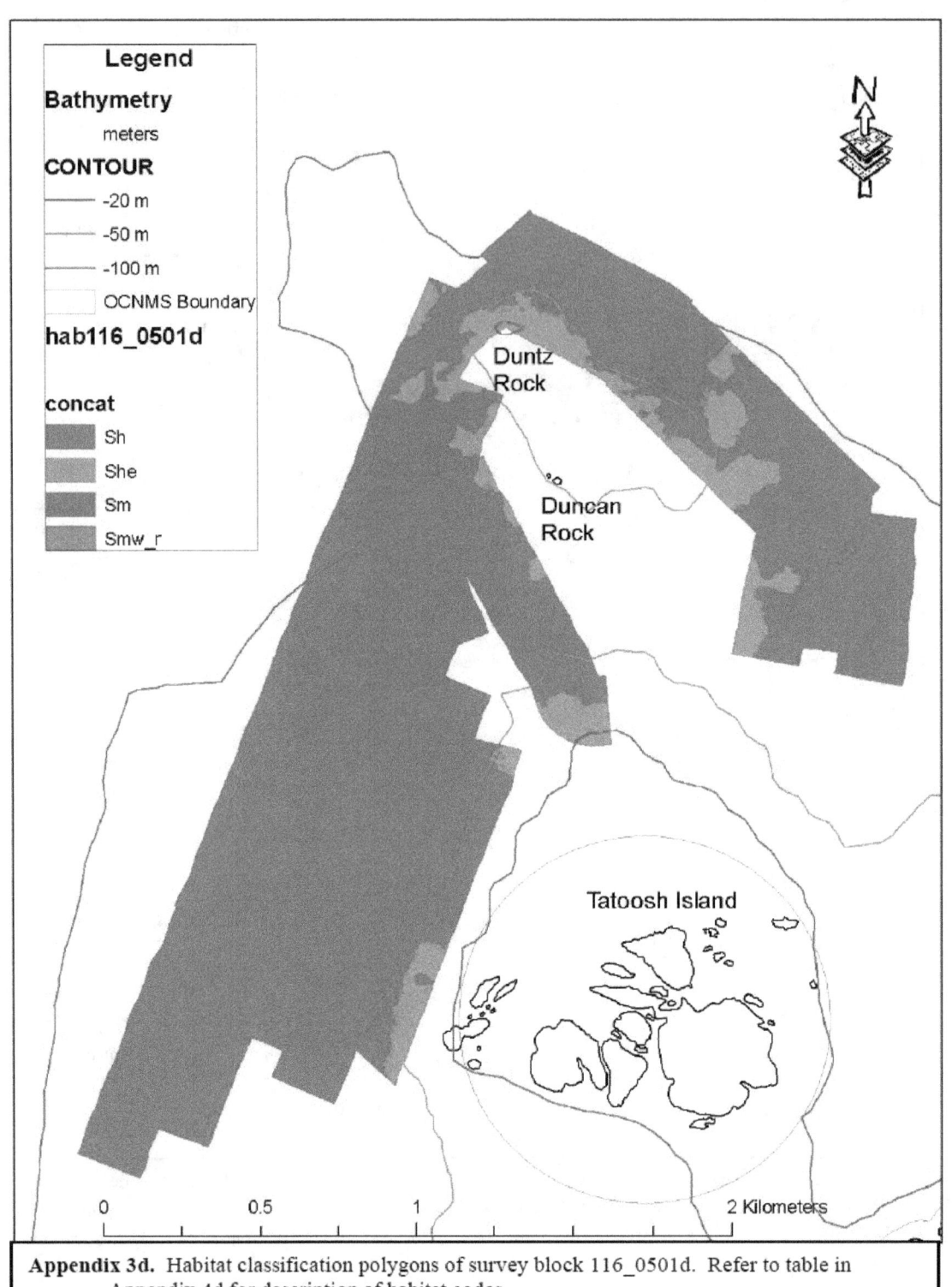

**Legend**

**Bathymetry**

meters

**CONTOUR**

—— -20 m

—— -50 m

—— -100 m

OCNMS Boundary

**hab116_0501d**

**concat**

Sh

She

Sm

Smw_r

Duntz Rock

Duncan Rock

Tatoosh Island

0   0.5   1   2 Kilometers

**Appendix 3d.** Habitat classification polygons of survey block 116_0501d. Refer to table in Appendix 4d for description of habitat codes.

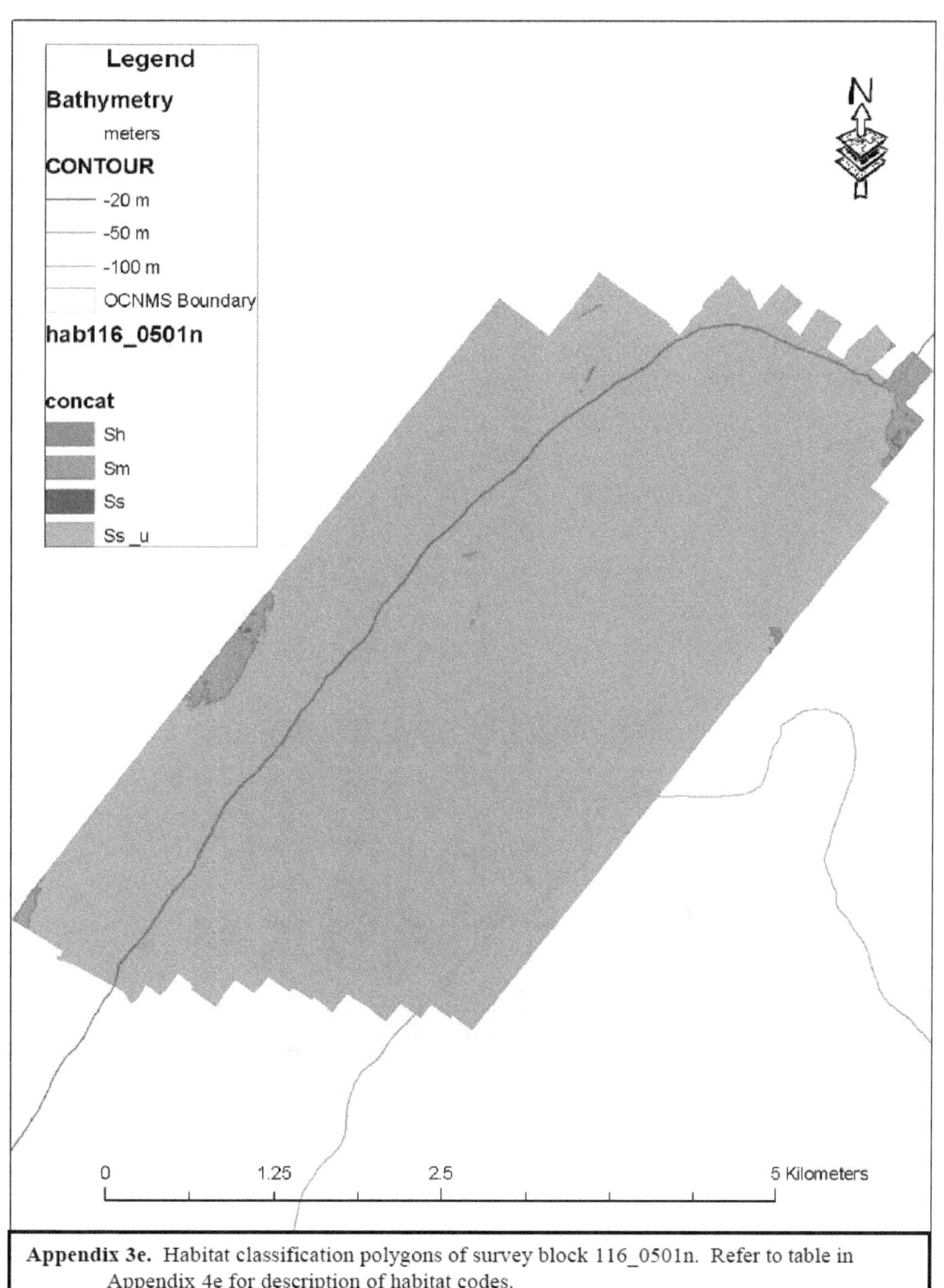

**Appendix 3e.** Habitat classification polygons of survey block 116_0501n. Refer to table in Appendix 4e for description of habitat codes.

29

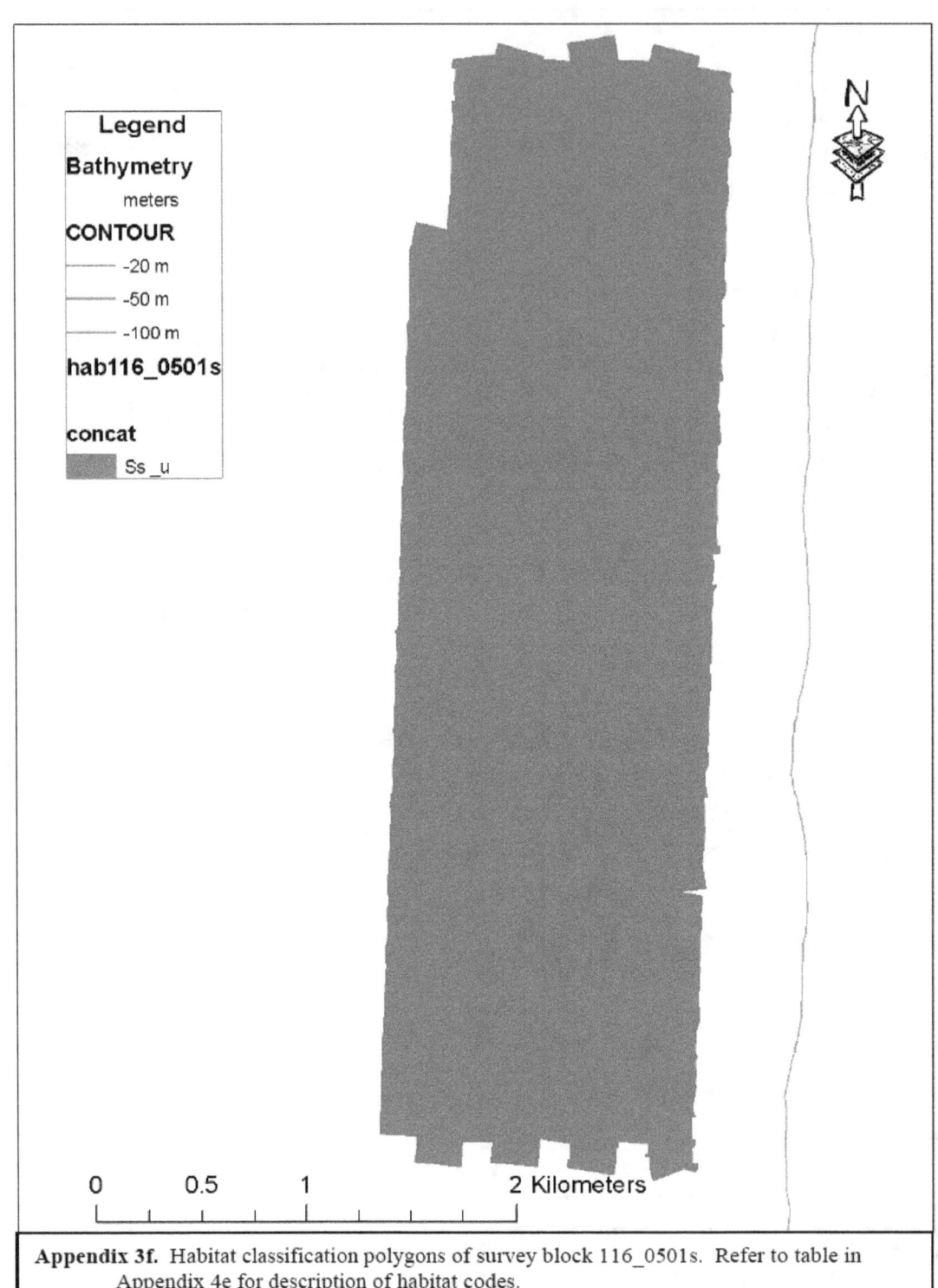

**Legend**

**Bathymetry**

meters

**CONTOUR**

—— -20 m

—— -50 m

—— -100 m

**hab116_0501s**

**concat**

Ss _u

0    0.5    1    2 Kilometers

**Appendix 3f.** Habitat classification polygons of survey block 116_0501s. Refer to table in Appendix 4e for description of habitat codes.

30

## Appendix 4. Habitat Classification Tables

**Appendix 4a.** Distribution of habitat classified from survey block 114_0402b side scan sonar survey data. Habitat codes are provided per Greene et al. (1999) and are presented by area in square meters, and area as a percentage of total mapped area.

| Habitat Code | Descriptor | Square m | Percentage |
|---|---|---|---|
| Ss_u | Shelf soft unconsolidated | 8,701,399 | 44.07 |
| Sm | Shelf mixed | 6,857,287 | 34.73 |
| Shd_d | Shelf hard deformed, differentially eroded | 1,557,155 | 7.89 |
| She | Shelf hard exposed | 1,495,263 | 7.57 |
| Sh | Shelf hard | 408,980 | 2.07 |
| Smw_r | Shelf mixed waves ripples | 286,531 | 1.45 |
| Sm_h | Shelf mixed hummocky | 227,932 | 1.15 |
| Ss | Shelf soft | 116,263 | 0.59 |
| Sh_d | Shelf hard differentially eroded | 91,695 | 0.46 |

**Appendix 4b.** Distribution of habitat classified from survey block 114_0402c side scan sonar survey data. Habitat codes are provided per Greene et al. (1999) and are presented by area in square meters, and area as a percentage of total mapped area.

| Habitat Code | Descriptor | Square m | Percentage |
|---|---|---|---|
| Sm | Shelf mixed | 1,001,959 | 79.02 |
| She | Shelf hard exposed | 252,934 | 19.95 |
| Sh | Shelf hard | 13,132 | 1.03 |

**Appendix 4c.** Distribution of habitat classified from survey block 116_0501c side scan sonar survey data. Habitat codes are provided per Greene et al. (1999) and are presented by area in square meters, and area as a percentage of total mapped area.

| Habitat Code | Descriptor | Square m | Percentage |
|---|---|---|---|
| Sm | Shelf mixed | 2,503,939 | 77.23 |
| She | Shelf hard exposed | 738,182 | 22.77 |

**Appendix 4d.** Distribution of habitat classified from survey block 116_0501d side scan sonar survey data. Habitat codes are provided per Greene et al. (1999) and are presented by area in square meters, and area as a percentage of total mapped area.

| Habitat Code | Descriptor | Square m | Percentage |
|---|---|---|---|
| Ss_u | Shelf soft unconsolidated | 19,829,820 | 98.20 |
| Sm | Shelf mixed | 341,667 | 1.69 |
| Ss | Shelf soft | 12,181 | 0.06 |
| Sh | Shelf hard | 10,240 | 0.05 |

**Appendix 4e.** Distribution of habitat classified from survey block 116_0501s side scan sonar survey data. Habitat codes are provided per Greene et al. (1999) and are presented by area in square meters, and area as a percentage of total mapped area.

| Habitat Code | Descriptor | Square m | Percentage |
|---|---|---|---|
| Ss_u | Shelf soft unconsolidated | 7,585,526 | 100 |

**Appendix 5. Groundtruthing images representative of associated habitat classes.**

Soft (s)

Mixed (m)

Hard (h)

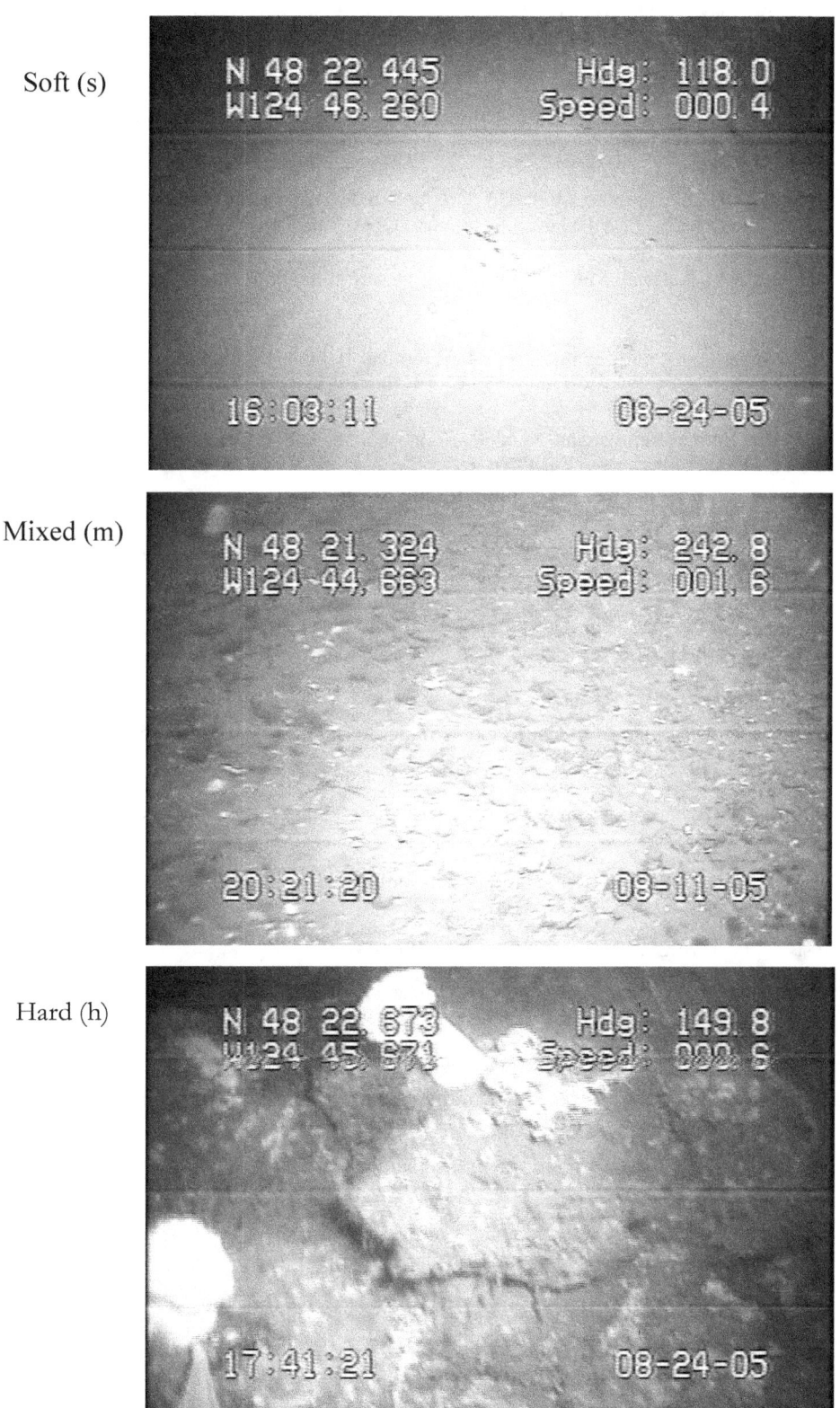

# ONMS CONSERVATION SERIES PUBLICATIONS

To date, the following reports have been published in the Marine Sanctuaries Conservation Series. All publications are available on the National Marine Sanctuary Program website (http://www.sanctuaries.noaa.gov/).

A Pilot Study of Hogfish (*Lachnolaimus maximus* Walbaum 1792) Movement in the Conch Reef Research Only Area (Northern Florida Keys) (NMSP-06-06)

Comments on Hydrographic and Topographic LIDAR Acquisition and Merging with Multibeam Sounding Data Acquired in the Olympic Coast National Marine Sanctuary (ONMS-06-05)

Conservation Science in NOAA's National Marine Sanctuaries: Description and Recent Accomplishments (ONMS-06-04)

Normalization and characterization of multibeam backscatter: Koitlah Point to Point of the Arches, Olympic Coast National Marine Sanctuary - Survey HMPR-115-2004-03 (ONMS-06-03)

Developing Alternatives for Optimal Representation of Seafloor Habitats and Associated Communities in Stellwagen Bank National Marine Sanctuary (ONMS-06-02)

Benthic Habitat Mapping in the Olympic Coast National Marine Sanctuary (ONMS-06-01)

Channel Islands Deep Water Monitoring Plan Development Workshop Report (ONMS-05-05)

Movement of yellowtail snapper (Ocyurus chrysurus Block 1790) and black grouper (Mycteroperca bonaci Poey 1860) in the northern Florida Keys National Marine Sanctuary as determined by acoustic telemetry (MSD-05-4)

The Impacts of Coastal Protection Structures in California's Monterey Bay National Marine Sanctuary (MSD-05-3)

An annotated bibliography of diet studies of fish of the southeast United States and Gray's Reef National Marine Sanctuary (MSD-05-2)

Noise Levels and Sources in the Stellwagen Bank National Marine Sanctuary and the St. Lawrence River Estuary (MSD-05-1)

Biogeographic Analysis of the Tortugas Ecological Reserve (MSD-04-1)

A Review of the Ecological Effectiveness of Subtidal Marine Reserves in Central California (MSD-04-2, MSD-04-3)

Pre-Construction Coral Survey of the M/V Wellwood Grounding Site (MSD-03-1)

Olympic Coast National Marine Sanctuary: Proceedings of the 1998 Research Workshop, Seattle, Washington (MSD-01-04)

Workshop on Marine Mammal Research & Monitoring in the National Marine Sanctuaries (MSD-01-03)

A Review of Marine Zones in the Monterey Bay National Marine Sanctuary (MSD-01-2)

Distribution and Sighting Frequency of Reef Fishes in the Florida Keys National Marine Sanctuary (MSD-01-1)

Flower Garden Banks National Marine Sanctuary: A Rapid Assessment of Coral, Fish, and Algae Using the AGRRA Protocol (MSD-00-3)

The Economic Contribution of Whalewatching to Regional Economies: Perspectives From Two National Marine Sanctuaries (MSD-00-2)

Olympic Coast National Marine Sanctuary Area to be Avoided Education and Monitoring Program (MSD-00-1)

Multi-species and Multi-interest Management: an Ecosystem Approach to Market Squid (*Loligo opalescens*) Harvest in California (MSD-99-1)

www.ingramcontent.com/pod-product-compliance
Lightning Source LLC
Chambersburg PA
CBHW081802170526
45167CB00008B/3287